The
Evolution
of an
Independent
Home

TRADEMARKS

The
Evolution
of an
Independent
Home

The Story of a
Solar Electric Pioneer

By PAUL JEFFREY FOWLER

FOWLER ENTERPRISES
Worthington, Massachusetts
1995

Copyright © 1995 by Paul Jeffrey Fowler

Printed in the United States of America.

Fowler, Paul Jeffrey
    The Evolution of an Independent Home: The Story of a
Solar Electric Pioneer / Paul Jeffrey Fowler.
      p.   cm.
    Includes index.
    ISBN 0-9645111-7-7 (paperback)
    1. Dwellings—Energy Conservation. 2. Renewable energy sources.
I. Title.
    Library of Congress Catalog Card Number: 95-90039
    696—dc20

Published by:
Fowler Enterprises
264 Bashan Hill Road
P.O. Box 253
Worthington, MA 01098-0253

Cover photograph by Paul Jeffrey Fowler
Book and cover design by Jeff Potter

10 9 8 7 6 5 4 3 2 1

Due to variability of local conditions, materials, skills, site, and so forth, Fowler Enterprises and the author assume no responsibility for personal injury, property damage, or loss from actions inspired by information in this book. Always consult the manufacturer, applicable building codes, and the National Electrical Code® before installing or operating home energy systems. For systems that will be connected to the utility grid, always check with your local utility first. When in doubt, ask for advice; recommendations in this book are no substitute for the directives of equipment manufacturers or federal, state, and local regulatory agencies.

TABLE OF CONTENTS

ILLUSTRATION CREDITS

WARNING —
DISCLAIMER

THIS BOOK is the autobiographical account of building this specific independent home. This book is not meant to instruct anyone in building, designing, or wiring a home or a solar electric system. The drawings, photographs, and text are insufficiently detailed to use as guidelines for the described designs, construction, or installations.

If you wish to build or modify a home or solar electric system you should seek appropriate technical information and the services of competent professionals. Always consult the manufacturers, applicable building codes, and the National Electrical Code® before building, wiring, or altering houses or installing or operating home energy systems.

Every effort has been made to make this book as complete and accurate as possible. However, there **may be mistakes**, both typographical and in content. Therefore, this text, photographs, and drawings should only be used as a general guide to tell the story.

The purpose of this book is to educate and entertain. The author and Fowler Enterprises shall have neither liability nor responsibility to any person or entity with respect to any loss or damage caused, or alleged to be caused, directly or indirectly by the information contained in this book.

If you do not wish to be bound by the above, you may return this book to the publisher for a full refund.

INTRODUCTION
& ACKNOWLEDGMENTS

MOST OF ALL I would like the readers to know how much I enjoyed this project. I managed to write and illustrate this book on a schedule that sacrificed neither my family time nor our independent way of life. I worked on the book from 7 A.M. to noon on weekday mornings. In the afternoons I cared for our two-year-old son, Terry and worked on our homestead with my wife, Lea. On weekends our life revolved around my three stepchildren: Kurt, Bethy, and Jarod.

Building this house and homestead has been a long and sometimes demanding project. In the early years I had to do too much by myself. In later years it was been a challenge to find time while running a solar electric business and settling in with the needs of my new family. There are still shelves that need to be installed in closets and mopboards that have yet to be nailed in place. There are lists of new projects to be undertaken.

The writing of this book has completed my independent home for me emotionally. As I have sat and written each day, I have reanalyzed my many decisions. I am now content with them from beginning to end. I have reflected on the poor choices I made and the mistakes that were made by others who worked for me on both the house project and for my business. There is harmony in my independent home project. I feel this book is the best thing I have ever done. It is more important and more satisfying to me than the technical books I wrote, the house structure I built, or the business I created and later sold.

I portray the actions of most of the people in a very positive manner because they were good people with whom I thoroughly enjoyed working and living in my small town. As in any project, I had my share of difficult situations. A few of these are in this book to tell the whole story properly. I have taken the liberty to disguise or leave out the names in these parts. These people's lives are more important than my anecdotes. Also, I relate my stories from my memory. I wish they could be 100 percent accurate, though I am sure they are not. If any reader remembers events differently, I sincerely apologize.

* * * * *

THE PRODUCTION of this book has been accomplished with local hilltown talent. Western Massachusetts and the Berkshire Hills in particular abound with accomplished individuals who choose to live in these beautiful surroundings though professional opportunities are limited. Consequently, I have had many talented people to work with. Most of these people were my friends in the beginning; the rest have become my friends.

My editors included, my wife, Leatrice Fowler; my friend Sue Dunham; my friend and grammar school and high school teacher, Mrs. Ida Joslyn; my former assistant at Fowler Solar Electric Inc., Ron Woodland, and professional editor Jeff Potter. Each provided a different insight for the development of my manuscript. My most emotional thank you goes to my wife, Lea, who labored through my first and roughest draft, all the while encouraging me to go on.

My most professional thank you goes to Jeff Potter. I had the vision, but Jeff had the expertise to put all the components together. Jeff scanned and digitally enhanced the photographs and designed the book in addition to being the chief and final copy editor.

I originally started writing this book for myself. I think I wanted to write this book from the first day I took possession of my remote land on Bashan Hill. As the project wore on and the tasks became more difficult, I finished the book for my family: Terry, Lea, Kurt, Bethy, and Jarod. Eventually in the next twenty years, each will read this book and better understand how and where they have lived.

—*Jeffrey Fowler*

CHAPTER 1

My Roots in Self-Sufficiency

I T HAS ALWAYS BEEN difficult for me to comprehend where things begin and where things end. In this case, it is hard for me to tell you exactly where my life in a solar electric home on an environmentally sound homestead actually began. Sometimes it seems it began on the day I sent my check to Joel Davidson to be part of a cooperative buy of solar electric modules. Most of the time, I believe this way of life began long before it actually did begin. Sometimes I think it started with my grandfather, John Stowe Fowler.

I never met any of my grandparents. My Grandfather Hill died when my mother was ten, and my mother grew up with two aunts and an uncle and saw her mother only a few times, the last time when she was eighteen. Grandfather Fowler died the same year I was born. Grandmother Fowler lived in California until I was sixteen, light years away because my father wanted to be separate from his remaining family.

My father told me long stories about his childhood and Grandfather Fowler. I would often ask when Grandfather Fowler had died. I remember plainly at the age of four thinking I was my grandfather after he had died. Looking back as an adult I know that even as a precocious young child, I had not been taught what reincarnation was. I probably had heard a vague reference to the idea in a grownup's conversation. For whatever reason or reality, I felt that part of me was my grandfather because he had died three months before I was born. I was smart enough, or lucky enough, never to talk much with my parents about strange thoughts such as being

My grandfather John Stowe Fowler at the age of 20 in 1902.

connected with my grandfather through some idea of reincarnation.

I think a good deal about my childhood and childhood development these days. Most of my thoughts revolve around my two-year-old son, Terry, my wife, Lea, and my three stepchildren, Jarod, Beth, and Kurt. These things were probably inevitable when I became a stepfather at the age of 42 and a father at the age of 44.

I am not involved in the occult world, but I do still feel a connection to my grandfather. I wish I had known John Fowler, a very good man and a good father. Unlike me, he was not an educated man. He grew up about 40 miles north of where I grew up, in a remarkably similar Colonial house, but in southern Vermont. I think my way of life began with him, when I remember the stories my father told of how he could build or fix anything and how he grew tired of the monopoly of the local power plant and disconnected his garage and apartments from the power company. In 1918, he saved the oil from oil changes in the garage he owned to fuel a system to power the whole building.

My father died June 30, 1978, five months after I turned 30 on February 1. I went to visit my father in my hometown of Worthington, Massachusetts, the week of my birthday. I had never told him my thoughts of my connection to my grandfather. This time, when our conversation passed to his childhood and his father, an idea flashed into my mind. I asked, "When was Grandfather Fowler born?" My father replied, "He was born on January 31."

My grandfather and I were both born under the sign of Aquarius. The descriptions for this sign are filled with phrases such as, "fifty years ahead of the times," "inventive," "scientific with an emphasis on electricity," or "sun sign of genius." So, some days I feel my way of life really began with my grandfather.

MY FIRST ECONOMIC COMMITMENT to a solar electric way of life came on May 18, 1981. I simultaneously sold my two homes on the other end of town on six acres of land (left over from the farm on which I had grown up), and took possession of nine acres of land — 1.3 miles and $20,000 removed from the nearest power line.

It was no accident that I acquired a piece of land that had no power to it. I wanted what I was getting and had been searching it out for many years. During subsequent years with my business, Fowler Solar Electric Inc., I met many people who bought a piece of land only to realize later that the power company would not come to their new home site for free. I was the opposite.

In 1978, a few months after my father died, I went to Europe with a few dollars and a bag of clothes over my shoulder. I met a woman who was an art student, and I stayed in London that winter. I was alone most of the time in a dreadfully cold flat. I had no friends. I very much did not fit into the stiff-upper-lip English world. My days consisted of getting warm enough to play guitar for an hour and then retreating under the down comforter for the rest of the non-eating hours to read. I held the book in one hand while warming the other hand under the covers. When the book hand got numb I would switch.

One day I took a day trip to the other side of London to the largest bookstore I had ever seen. It was like a department store with nothing but books. I was browsing in a section on house design and building when I found a book titled *Technological Self-Sufficiency*, by Robin Clarke. I bought it because it discussed designing, building, and heating a homestead. Its emphasis was on knowledge and science and new environmental ways to do it. The second half of the book was on solar heating and making your own electricity from wind power, hydropower, and methane. I read the book and then slipped into my daydreaming and designing mode. From that day on I have probably spent part of most any day thinking about designing or upgrading my home, car, or power system.

IN THE FALL OF 1979, I bought the house down the road from the mobile home I had inherited from my father. Ostensibly, I bought it because I owned the land on both sides of it as part of the land I had inherited with my father's trailer. However, the real reason I bought it was that I hated that trailer.

I had grown up in Worthington on the 120 acre farm just up the hill. Thirty acres of fields were divided by stone walls and dotted with a chicken house, ice house, tool shed, maple sugar house, milk house, corn crib, a large barn, and a large Colonial home with a large ell. We moved there from the other side of town when I was four. To me, it was my place in the world. My father was not a farmer, but an executive in a small company in the city that was 20 miles away. In his absence I was the king of all the lands and all the buildings. I was raised with the promise that this place would be mine someday. I could write a book in itself about the mountain brook, the valley river, the maple woods, pine grove, and two drumlins covered with blueberry bushes.

My father was fired from his job in the summer before my senior year of college when the little company he worked for was swallowed up by a big company. I had paid half of my way through an expensive private school for the first three years of college, and I was broke. Since I was a very independent 21-year-old counterculture student in 1969, and since my father was a conservative 60-year-old man with a bad drinking problem, there was no family money or loan for any part of my last year of college. The school and the banks would not lend me money because my father had money. My promised farm was sold six months later.

My mother would not let my father build on a piece of land from the edge

Ten-year-old Jeffrey Fowler in front of his Colonial farmhouse in 1958.

of the old farm because she knew he would never finish it. The compromise was the large mobile home, where she died two years later.

When my father died six years after my mother, I inherited the mobile home and a few acres of land. I left my attachments to the farm behind me and moved from Cambridge (next to Boston) back to my little hometown in the Berkshires and cleaned up the trailer and the property. I have always felt I could make something out of nothing. I managed to get rid of the evidence of my father having lived there for six years by himself without cleaning. But no matter how long I thought about redesigning the trailer, I came back to the same realization: there was nothing worth saving. The windows were poorly made, the insulation was insufficient, and the kitchen cabinets were cheap.

So IN 1979, I bought my sister's house next door. I really wanted to design and build a house, but I did not have the money. My sister had recently divorced, remarried, and relocated in the next town. She needed badly to sell her house at a time when no one wanted to buy a house in Worthington. I knew the house from the inside out. My friend Tom Whalen and I had built the house five years previously. The price was good. She waited until the real estate agreement ran out and sold it to me minus the commission. It was the safe thing for me to do.

Tom was my best friend in 1974. We were both carpenters. I was very close to my sister, her then husband, Kenny Beach, and their two children. I was living in a cabin with no utilities which I had built on the back of their land. Tom became best friends with them, too. Sue and Kenny were crowded in a small trailer without the knowhow to build their own house or the money to have someone build it.

After a lot of maneuvering with the bank, we came up with a plan. Sue and Kenny would get a loan for materials to build their own small Cape Cod–style house next to their trailer. Tom and I would frame the house for almost free, and they would finish the inside. Tom and I, along with Sue and her ten-year-old son Scott, framed the house in two weeks. Kenny worked on the house evenings and weekends. Sue, Scott, and I later put up Sheetrock in the house in a week. Later, I started Kenny on the flooring and siding.

This house was simple, basic, and partly unfinished, but it had no big problems. I spent the whole time that I lived there wishing I could redesign it to make it into something more than it was. For Christmas I had received a subscription to *Solar Age* magazine. I bought books on passive solar design and alternative energy. I redesigned that poor house a thousand ways, attempting to transform it into a passive solar efficient home. All I accomplished was to educate myself. At the end of my course, I realized that there was no way I could remake that house into a passive solar home without moving it from the river valley in which it was built. The solar exposure was just too little.

I became friends with Jonathan Ginzberg, the acupuncturist in town. We had somewhat similar backgrounds in that we had both attended Tufts University and we both were studying *Solar Age* magazine. He was renting a house in Worthington and planned to buy a piece of land and build a passive solar home. I started going along with him to look at land. We considered finding the perfect 40 acre piece of land that we could cut in two. Of course this never really happened, but the exercise did get me out of my house and familiar with the local real estate agents.

Once I got some confidence, I followed the land advertisements and talked to the real estate agents. The agents were either a little lazy about showing land that would not yield much of a commission, or they thought I was an unlikely buyer chewing up their time. They always gave me directions to some backwoods property and told me to look for myself. I never saw one decent piece of land that had the one basic characteristic that I required: south-facing solar exposure. I did have one other basic requirement at the time. The land could not be near any potato fields. This was a tough requirement since Albert Farms cultivated the majority of the farmland in our town in potatoes. I had grown up as a friend of the Albert family. However, for the previous two years, I had waged a bitter fight, along with some other people in town, to get Albert Farms to stop the aerial spraying of their fields, and correspondingly the spraying of the land of their abutters. As I learned more about the chemicals, I decided I could only own land upstream and distant from the fields.

I started taking topographical maps with me every time I went to a real estate agent. I found I could save a lot of mileage and wasted time by orienting the plot plan of the land for sale on the topo map; this would show if a mountain was in the way of solar exposure. Next I shaded in all of the Albert Farms potato fields. I could estimate the distance from the nearest field. By following the contours of the slopes and river valleys, I could estimate the direction of field runoff and the direction of the flow of the potential underground plume of water pollution.

BY THIS POINT in the project, I had an agreement and closing date to sell my father's trailer. I had advertised it in the local paper without listing it with a real estate agent. I got a response in a few days from an engineer in Worthington who was looking for an entry level home for his daughter and family. I showed him all the positives and negatives. We negotiated a price of $18,000, a price that was only $5000 above the value of the land, well, and the septic system. They planned to build a house and foundation around the trailer and turn it into a house, which they did a few years later.

My pie-in-the-sky game of planning, plotting, and looking for land for a solar house was becoming a reality. I actually had the money coming in from the sale of the trailer and could purchase a piece of land. If I found a piece of land, I could then plan to complete some of the unfinished parts of my current house and put it up for sale. I hoped to be able to scratch out enough money and time in that year to have a foundation on the new piece of land before I would have to move. The wheels of change turned more quickly and definitively than I would have guessed; in fact, once spinning quickly, they revved out of control. Within two months, I had sold both my house and trailer and taken possession of the piece of land with good solar exposure that was far from the power lines. In another three months I was living in the unfinished shell of my self-built passive solar home.

CHAPTER 2

The Road to Bashan Hill

MARY JANE CAREY was the new owner of Corners Realty in Worthington. She was by far the nicest and best real estate agent I had met. However, like most agents in 1981, she did not yet know the value of solar exposure, nor did she really understand what it was. She had sent me to see several pieces of land that were supposed to have had southern exposure, but did not. Since I liked her the most, I had made another appointment with her at her office for a Saturday in the first week of March.

I took my topographical maps. I showed her specific parts of roads in specific parts of town that did have solar exposure and were not threatened by groundwater pollution now, or likely to be threatened in the future. I then asked her what she had for listings in my designated areas. She had one listing for a double lot of nine acres on Bashan Hill Road. Then she dropped the clincher. "I never had anyone look at this piece because the power line is 1.3 miles away. It would cost a fortune to have the power company bring in the poles," she explained.

I surprised Mary Jane when I asked to see the land. She was not aware that I had lived in a cabin with no utilities for a few years and that my recent best friends in Worthington, Richard and Meg, lived in a cottage removed from the power lines. All the information from all the books about wind machines and independent power surged in my brain.

Worthington is located in the eastern part of the Berkshire Hills. Bashan Hill Road climbs within 50 feet of altitude of the top of this

Steep ascent of the
beginning of
Bashan Hill Road.

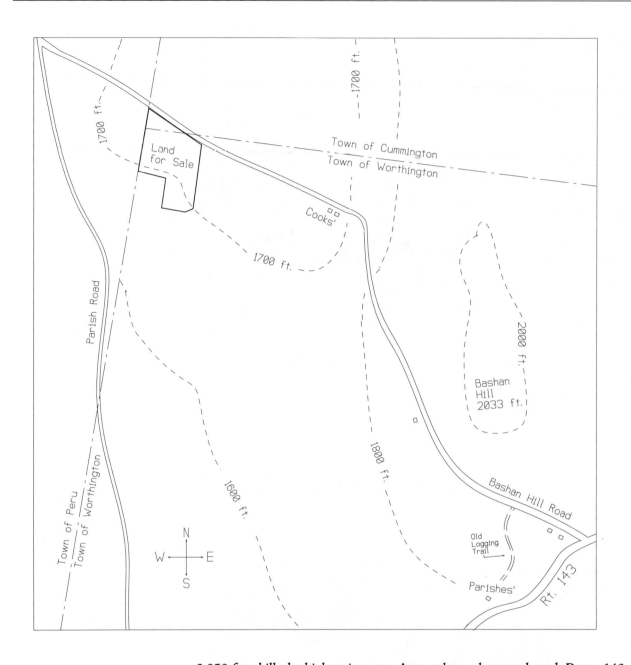

Map of Bashan Hill Road

2,050-foot hill, the highest in town. As you leave the paved road, Route 143, and turn onto Bashan Hill Road, you drive two tenths of a mile on a flat. Then you encounter a true billy-goat hill. The road levels off on a high flat on the west side of the hill. Now that I have lived here for many years I know this flat is entered by a snowmobile trail, part of which is actually an old logging road that extends to Sharon Parish Guy's house back on Route 143.

I knew Bashan Hill Road from growing up in Worthington. Every "all-boy type kid" like myself from a small town spent many Saturday nights driving every passable back road in town. My friends and I would have a few beers hidden in a river to stay cold. We would log fifty miles on a car, each have one beer, and pretend we were big drinkers.

My first memory of Bashan Hill Road was an adventure with Butch Parish and my sister when I was 9 years old. Butch was Sharon's older brother, about three years older than I. We were at the Parishes' on a Sunday afternoon. Butch had what we called a jalopy, an old Model A Ford that had the rear body cut off and replaced with a sort of wooden flatbed. Many of my friends had an old car or pickup that they and their fathers had gotten running just so they could drive it around the fields on their farms. Butch was told that he could take us for a ride around the backyard. My sister Sue was thirteen and pretty, so she got to ride in the front. I was the younger brother and got the back. As soon as we became sure all the parents were inside drinking, we headed up the woods road to Bashan Hill Road. Butch got the old Ford going as fast as he could down the flat. To impress us, he got all four wheels off the ground when he raced over the crest of the hill that descended steeply down Bashan Hill Road. I went flying in the back of the flatbed. I managed to stay in the truck, but I badly scraped my whole back. My sister, always worried for my welfare, told me to put my shirt on and to not let my mother see my back for a week so we would not get caught.

I LEFT CORNERS REALTY midmorning that Saturday for Bashan Hill Road. When I turned onto the road, I found the snowplow had not yet been there. I drove through five inches of new snow in a nearly new front-wheel-drive Dodge Colt with nice new Michelin tires. This car had gone anywhere all winter long, but it would not go up this steep one tenth of a mile. After three tries I backed up and left the car at the main road. I wrapped the dog leash around my waist and let my little dog Rolo run free.

There were no car tracks on the road. As I walked up the steep hill, I felt that I was walking on a path in the woods, not on a snow-covered road. After cresting at the top of the hill, the road continued at the same altitude for a third of a mile. This was not a plateau, but a flat area the width of the road along the side hill. I walked along for another quarter of a mile down a gentle grade. The road then dropped off quickly, took a 90° turn to the left, then went up a sharp baby hill and down again before I was on a plateau on the west side of Bashan Hill at 1,700 feet.

I was very glad I had not managed to make it up the first hill with my little car. If I had, I would have been on the inside of the valley trying to get out until a snowplow arrived. As I faced back up the hill, I could see that the trip out would always be worse than the trip in. The 90° corner would cause my car to lose the momentum that it would need to climb a snow-covered hill.

While I was enjoying the walk in, I was also trying to analyze the lay of the land. I was skeptical about the quality of the land I was going to look at, as I surveyed the environment in the first mile. There had been nothing but steep

At the bottom of Bashan Hill was a little brown house at the beginning of a level area.

grades on each side of the road with obvious signs of bedrock everywhere. The land was not suitable for a home, septic system, or driveway. The only structure was an old camp hanging off the side hill. Later I learned that it was basically cemented on a 45° slope of bare bedrock. It was obvious to me that no one would ever be developing the first mile of Bashan Hill Road. It also became clear that the power line would not come down Bashan Hill Road one house at a time and arrive at my prospective piece of land for free.

At exactly one mile, Bashan Hill Road straightened out on a flat in what felt like a valley after coming off the hill. It was actually a flat area before the next big hill a half mile down the road in the neighboring town of Peru. On the left side of the road sat a little old brown house with a garage next to it. It was obvious that someone lived there year round. The real estate agent had told me that Bob and Karin Cook lived here. I had met their son at a folk music event several years back. My sister and her husband had met them and had been friendly with them. Kenny and Sue had been impressed with Bob's guitar playing. I liked the idea of having a good guitar player next door.

That first day (and every time I drive home even now) I was impressed by the beauty of Bashan Hill Road. The parcel of land for sale began three-tenths of a mile beyond the Cooks' house. The road was flanked on both sides by large old maple trees and stone walls. These maples were beautiful even in the first week of March as they stood naked of their leaves. I found the boundary, and Rolo and I left the road to explore the land.

The land was on the left side, and the south side of the road. Beyond the stone wall and the maples were 25 foot Christmas trees. They had obviously been planted in the fields of an old farm. No one had tended them or harvested them at the correct size and now they were too tall. They would be a

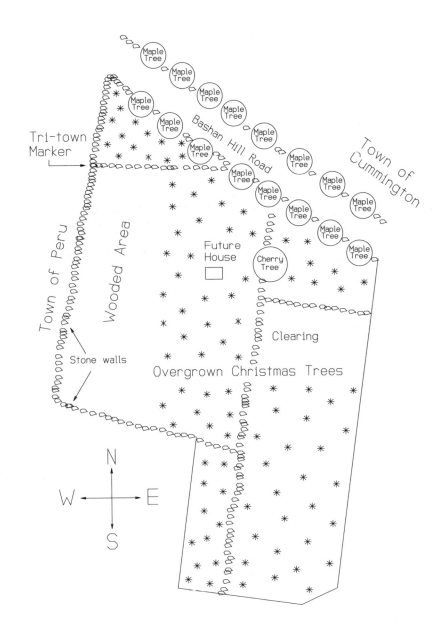

Map of the nine acres of land for sale on Bashan Hill Road.

Great Wall of China from the rest of the world. About two-hundred feet into the land two stone walls intersected in a small clearing. It was simply beautiful. The land was good and the terrain was flat, unlike much of the land for sale in these hilltowns. To the south, the land gently sloped away.

As I walked into the snow-covered clearing, a flock of yellow grosbeaks swooped in unison and landed simultaneously in an old apple tree. Rolo arrived from his own independent investigation. As he approached the apple tree, the grosbeaks made a great display of taking off in unison and landing in a more distant apple tree. This was heaven. Privacy, no cars, just birds and freshly fallen snow. My little dog was another reason I wanted to leave

Mature maple trees lined the road along the frontage of the land for sale.

my sister's old house in the valley. The cars speeding past the house were an ever-present danger for Rolo. In many ways this would be a retirement home for him for the last five years of his life.

When I had walked down the hill, I had felt I was in a valley. Now, after surveying the lot I realized I was on a plateau. I could look out in all directions. To the west I could look up at Bashan Hill covered in its fresh snow. At the edge of the clearing stood a large cherry tree. I love to climb trees. Some childhood loves never go away. I climbed this cherry tree to see if I might have a view from the upstairs of a house. I climbed to about the height of the top of a second story window. It was not a postcard picture-perfect descending view that New York summer folk love, but I could see Knob Hill in Otis, 10 miles away. I also could see no houses and no towers in any direction. This was a location where I could look in any direction in the night sky and not see neighbors.

I HAD SEEN ENOUGH of the land to know that this parcel was probably the place. It was certainly enough of the right place to start an in-depth final analysis. Rolo and I retraced our tracks in the snow back to the little brown house to meet the people who would later become my neighbors. I left Rolo outside to play with the Cooks' female black lab. I knocked at the door. I introduced myself, and two very tall people invited me to enter the little house.

Most people think I am tall. I am slightly over six feet in my bare feet. I appear to be taller because I am a rangy basketball-player type, and I have the very low body fat of a long time non-meateater. Karin Cook greeted me at the door and looked almost straight into my eyes. She was only about an inch shorter than I. Bob stood up from the table, walked over, reached down, and shook my hand. He was six-foot four and looked as if he had shoulder pads on under his chamois shirt. I later learned they had a son in high school who stood six-foot three, and another son the same size who lived in Boston. I reassessed my future as the little guy in the valley of the giants.

As I look back, I see that this was the perfect time to meet my new neighbors. It was too cold to boil sap that day, and it was the end of a long winter for people who had a touch of cabin fever. Several people had looked at parcels of land on Bashan Hill Road during the ten years the Cooks had lived there. As we talked, I felt they looked at me as the first one to fit the mold. We talked energetically for an hour until lunchtime, when I politely said I had to leave. I certainly did not want to overstay my welcome the first day.

Bob and Karin explained that the land I was looking at was a piece of Jewitt Stern's old farm. The land had been purchased by someone who had planted Christmas trees in the old fields. That owner had died before it was time to harvest them. They had then overgrown while the land was transferred to his heirs. The heirs wanted nothing to do with the land. Jewitt Stern had had a little old house and barn that were now two piles of rotten boards. Eventually the land was sold to a land developer, who hacked the farm into small lots.

When Bob and Karin had bought their land next to the old Stern property, Bob had researched the soil quality. It was a bit stony but good soil with good drainage underneath. Ironically, in the 1950s, Albert Farms had wanted to purchase the farm to grow potatoes. They knew the soil was great. They backed away because the road was so bad. They knew they would not be able to get their equipment in for spring planting. The Cooks had walked to their house during mud season for many years. Now the road was upgraded and solid with good gravel, but only to the Cooks' house. If I bought the land I would have to walk to my property from the Cooks' house during mud season. Maybe over several years I would be able to get the town to put in new gravel up to my land.

Soon the conversation turned to living with no power. Bob and Karin were not interested in paying to bring in a power line. They had a large Miller welder-generator for the washing machine and power tools. Their water flowed by gravity from a spring up on the hillside. This eliminated the whole problem of needing electricity to pump water. They were not happy with their kerosene lights, but these were only temporary until they installed an alternative energy system. Unfortunately, after ten years of living on Bashan Hill Road they did not yet have the time or money to get that far. Refrigeration was supplied by an old Servel propane refrigerator.

Bob told me how this was a good wind site. We then discussed wind generators, alternative energy systems, and passive solar energy design. We were both surprised to find we had mutual interests at similar levels of expertise.

As our visit came to a close, I was beginning to guess that I had just

Karin and Bob Cook's house.

scratched the surface of my prospective new neighbors. I knew Bob and Karin were homesteaders and maple sugar producers. Bob was a union iron-worker and construction worker, and Karin was a watercolor artist. My intuition told me that they would be another example of the Worthington new-comers who had impressive backgrounds somewhere in the past. A friend later told me that Professor Cook had taught him sociology at Yale.

I left with a promise to inform them of my progress with the land purchase. They offered help and information if I needed it. I gathered up my trusty dog Rolo and daydreamed as we walked out. I am sure Rolo daydreamed about the big, cute, female black lab Annie.

Once home, I tried to calm down from the exciting morning and organize my thoughts. I made a mental list of my needs for electricity and how I could supply them alternatively. In later years, I was always pleasantly surprised when I discovered how many people in the little towns of New England live on remote land with no power. At this time Worthington had its own selection of people who lived off the grid. I decided to consult my friends Richard and Meg and my longtime friend from childhood, Guy Mason, who owned homes independent of the power company.

Richard Mansfield and
Meg Breymann's home.

CHAPTER 3

A Quick Decision

RICHARD AND MEG belonged to the part of our town's population which the local conservatives referred to as "still fighting the anti-Vietnam War." They grouped me, of course, with Meg and Richard, even though I was a local boy, largely because I had lived other places during and after college. We all were interested in organic gardening, building our own homes, non-toxic ways of life, and, in general, not destroying the world. Most of us had been influenced by the sixties. I had short hair, but somehow there still remained a shadow of my old ponytail and a telltale gleam of defiance in my eyes.

One day in the summer of 1980, I had driven up Route 112 between two potato fields, only to be doused with fungicide as a spray plane oversprayed the field. This was distressing, but not the end of the world. I stopped at the general store, got the mail, bought a newspaper, and complained to a few friends about being sprayed. I took a non-confrontational approach and decided to alter my route out of town. I drove down Old Post Road to avoid the spray plane. Unfortunately, the plane had switched fields. It flew over my car on Old Post Road and sprayed me again.

This incident started a movement of people who protested Albert Farms' aerial spraying. Our movement that year set the stage for later town and state involvement when it was found that Worthington had groundwater pollution attributed to a pesticide that was used on the potato crop. Through the many meetings about the aerial spraying, I became friends with Richard and Meg.

Richard and Meg had a 40 acre piece of land on a back road in Worthington. At first, they started clearing land for a home near their border of an Albert Farms potato field. While they were working, the Albert Farms plane oversprayed the potato field and sprayed them. Richard and Meg moved their homesite as far from the field as possible. This new site was thousands of dollars from the power line.

After my first visit to my land on Bashan Hill Road, I phoned Richard to ask him what it was like to live off of the grid. He told me the most important need would be refrigeration. I would have to find one of the old Servel propane refrigerators that were popular in the forties and the fifties. He recommended gaslights over kerosene lights because gaslights burned cleaner and brighter. The house could be built with power tools powered by a small generator. In general, he found life a little harder but mainly different and more peaceful.

Richard had needed a telephone. The telephone company had not been able to install one in his home in the woods. He had finally resorted to getting a telephone service up at the road, and had then strung his own wires, 1,000 feet down through the woods.

Richard and Meg had one large problem that my future neighbors, the Cooks, did not have. They could have only a drilled well. The water level was 40 feet down in the deep well. Richard's solution was an old-time one. From a supply catalog out of West Virginia, he purchased an old-style, deep-well hand pump with a long, three-foot pump handle. The pump consisted of leather gaskets, or mini-pumps, mounted on a rod inside of a casing. The leather gasket pumped the water up six feet to the next leather gasket, to the next, to the next, until the water reached the top of the well.

The pump worked quite well, but Richard had to pump that handle whenever he wanted water. Richard had rigged up a 120 gallon storage tank in the attic of his cottage. He would go outside in all kinds of weather and sweat for twenty minutes each day to fill the tank. Later the tank gravity fed the house until the next pumping. An additional, and worse, problem, was the pump maintenance. The leather gaskets needed to be periodically replaced, which involved pulling 60 feet of rod and gaskets out of the casing and replacing the gaskets on the ground. Unfortunately, the well had often failed in the cold winter when the repair became extremely difficult.

GUY MASON BECAME MY FRIEND in the fifth grade when his family bought the Hicklings' farm on Kinnebrook Road. As kids, we had built forts, played in our respective hay barns, and had taken apart radios and stereos. When it came to electronics, I was the one who took apart the radios and had all the questions. It was Guy who had the knowledge to tell me how to put the radios together again. Guy went to the University of Massachusetts for one or two semesters. He hated the large and impersonal school, so he quit. He then educated himself, and still does today.

In adult life, Guy Robert Mason chose to be called "Bob" for his middle name. Bob and I were each 33 years old when I decided to consult with him about my homesite. We both were self-educated, highly evolved jacks-of-all-trades. Some of his strengths were electronics, carpentry, and mechanics. The best way to describe Bob's abilities is to say he could decide to do a project, and then go research it, design it, and build it with largely scavenged parts and materials.

The morning after I first visited the Bashan Hill property, I drove to Bob's house, which was on a piece of land that had been part of his family's large farm. Bob's land was on a distant, wooded corner on an old dirt road far from the power line. Bob hated to spend money for the power company to install poles when he could make his own power. He was pretty sure the power would come down in the future for free.

A large, six-foot-four man greeted me at the door, a broad friendly smile flashing out of his bushy red beard. It was, of course, Bob greeting me as he always had. Ever since I had met him on the first day of school in the fifth grade he had towered over me. Bob had built a house-sized work space and garage. He lived in a loft upstairs by himself. I had never been to his house before. I just loved the endless tools and parts of electronic and mechanical equipment. The building was built mostly from lumber he had logged from his own land.

Inside the garage work area of Bob's building stood a six-foot-long, very old generator that Bob had salvaged from the barn on his parents' farm. It had not been running for twenty years, so Bob made it run. The generator was powered by a very high quality Wakesha engine. The generator, unfortunately, was a 120VDC (120 volt direct current) model. It could not produce 120VAC (120 volt alternating current) electricity, which is the standard electricity of a home. Bob had hand picked or modified most of his tools so that they could be powered by the 120VDC.

Direct Current, or DC, is the necessary form of electricity for charging batteries. It is also the type of current that batteries give off to power appliances. For example, a 12V generator in your car produces 12VDC electricity to charge the 12V battery. The battery stores 12VDC electricity as chemical energy. The battery gives off 12VDC electricity to power the headlights when you turn them on.

Bob designed his electrical system to work similarly to a car's electrical system, except that it was configured to be a 120VDC system instead of a 12VDC car system. He had the 120VDC generator. A 120V battery was not available, so he wired ten 12V batteries in series to create one. All the loads in his building ran on 120VDC electricity that was supplied by the 120V battery bank.

The biggest problem of a generator being used to power a single home is that the nominal 4000 watt generator must run all the time you are powering even the smallest 15 watt load. This is not a problem for the electric company which supplies electricity constantly to a large number of homes and factories. For the single home, the noise of the generator becomes oppressive, the gasoline usage becomes expensive, and the generator wears out quickly because of the high number of hours of running time.

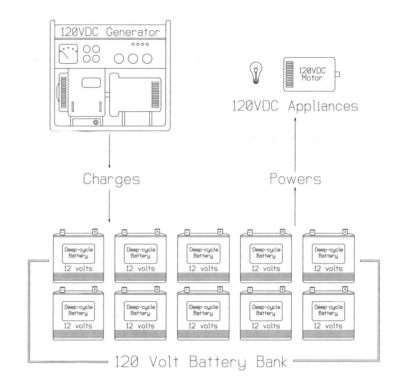

120VDC Generator

120VDC Motor

120VDC Appliances

Charges

Powers

Deep-cycle Battery 12 volts

Deep-cycle Battery 12 volts

Deep-cycle Battery 12 volts

Deep-cycle Battery 12 volts

Deep-cycle Battery 12 volts

Deep-cycle Battery 12 volts

Deep-cycle Battery 12 volts

Deep-cycle Battery 12 volts

Deep-cycle Battery 12 volts

Deep-cycle Battery 12 volts

120 Volt Battery Bank

Bob Mason's alternative electric system.

Bob had added the 120V battery bank to his generator to eliminate the need to run it each time he wanted to power a load. He would power the loads in his shop and home for a day by drawing electricity out of the battery bank. The next morning he would run the generator for an hour or so to return the batteries to full charge.

I spent Sunday afternoon organizing my thoughts on my new project. Speaking with Richard, Bob, and the Cooks had provided the "down to earth how do you live" component. I had many thoughts and ideas from reading books on alternative energy. But these were only ideas at this point. Each idea needed some testing and a good dose of trial and error. After speaking with the people who were actually doing it every day, I had the confidence that, if I never did anything more that just copy what they were doing, I could survive and be happy.

Sometimes people portray their actions in better than truthful terms as the length of time between the past event and the present increases. I like to think that I am being honest with myself, and accurate for you, when I say that at the time, I knew the alternative energy part of the project was way down the road. If I purchased this land, I would have to design the house, build the house, and create the garden before I could ever erect a wind machine or install a battery bank.

In that beginning of March of 1981, I did not have the responsibilities of a homestead. I was only working part time repairing guitars and banjos in my shop at home for a store in the city. I could concentrate my efforts on the decisions at hand.

My sensible friends would have advised me to take my time. I knew it made sense to look at the land many times. I could research different ideas over several months. The land had been for sale for a long time and was not about to be sold out from under me, but knowing all this, I did my best to avoid my sensible friends like the plague. I headed for deep water to see how well I might swim.

That Sunday afternoon I started to make lists of the pros and cons of the purchase. I was sure of the soil on the land from Bob Cook's information. I would need to double-check by looking at a soil map at the cooperative extension service. The land showed no outcropping of ledges. It had been perc tested before it was put up for sale. This meant that all was well with the Board of Health for the installation of a septic system. Additionally, since the perc test showed a fast perc rate, I knew the land had good subsoil drainage and little clay.

The land was on the south side of Bashan Hill Road, which ran east to west. This meant that the south-facing windows of my prospective passive solar house would look away from the road and toward the privacy of the Christmas tree lot. Nothing is much worse than living in a passive solar house facing the road, with the windows lit up at night like a giant drive-in movie for all to see.

There were no tall trees to the south of the prospective house site. Only a small amount of clearing would be necessary to allow the winter sun, which is low in the southern sky, to give full, all-day-long solar exposure to the south windows. Furthermore, the land, though relatively flat for the first 400 feet, sloped down to the south into a river valley at the end of this lot. This increased the quality of solar exposure. This land was an excellent solar site.

At this time, I planned to power my future alternative energy system with a wind machine. The prevailing winds come from the west and northwest in this area. The next ridge was a half-mile to the west. This ridge was only 300 feet higher in altitude than the prospective wind machine site. To the north, the slope was level, and to the south, the land sloped away. Even to the east, from where the prevailing winds do not blow, there was only Bashan Hill, 300 feet higher in altitude and a half mile away. The trees surrounding the house site were only about 30 feet tall. This was an excellent wind site. A 60 foot tower would put the wind machine the required minimum of 30 feet above obstructions. A 70 to 90 foot tower would catch the stronger winds.

Luckily, telephone service was not a problem on Bashan Hill Road. Telephone lines had been run to the Cooks', and then down past the land for sale, to a camp, at a time when the telephone company had not charged

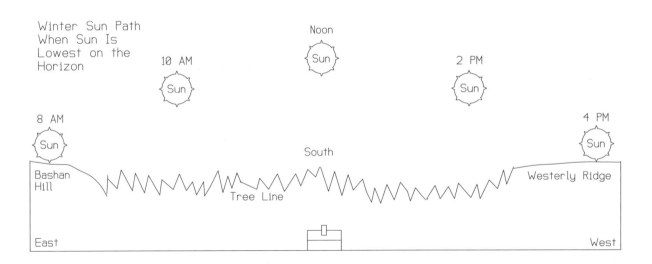

Winter Sun Path
When Sun Is
Lowest on the
Horizon

Noon
Sun

10 AM
Sun

2 PM
Sun

8 AM
Sun

4 PM
Sun

South

Bashan
Hill

Westerly Ridge

Tree Line

East

West

Solar exposure of prospective house site.

for line extensions. The telephone lines were not completely without problems. For a third of a mile on Bashan Hill Road, the telephone lines were hanging on standard insulators, attached to trees, not telephone poles. When the trees blew too much in the wind, the lines would fail. There was actually one line, with the two conductors inside, which ran one mile from the main road to the Cooks' home. A second line ran one and a half miles to the camp past the land at which I was looking. When I later ordered my phone, the telephone company ran a third line down the road to my land.

I had first visited the land while it was covered with snow. From the slope of the land in the general area, it was obvious that there could be no gravity feed water system. In fact, there would be no dependable shallow, dug well. The only choice would be a drilled, or deep, well. It would require a deep-well, submersible pump that would be powered by a generator. At this point, I did not even have a reasonable idea of how to power a deep-well pump from a wind system.

As I looked over my first list of the pros and cons of the project, I realized that most aspects looked good, very good. I could live with the lack of a gravity feed water system. Mainly, what I needed to do was double-check some of my data to make sure I was not making a decision on false assumptions.

I had basically made my decision to buy the land. I planned to visit the property once again the next morning. I would walk the boundaries and carefully plot out the potential house site with respect to the septic system, the well, and distances from the boundaries. Later in the day I would meet with Mary Jane Carey.

THERE WAS ONE OTHER PROBLEM: the price was too high. There were nine acres, and the people from New York were asking $12,500. Land at this time was selling for $1,000 per acre. This land was well over that figure and it did not even have access to a power line. Bob Cook had had a quote from the power company of $20,000 to bring the power line down to his house. In many respects, this land was really worth only about $7,500 as a piece of land suitable for a hunting cabin in the woods. The last third of a mile of road was impassable in the spring. Contrastingly, I was willing to pay a premium for a building lot with perfect solar exposure. At this time, in other parts of the country such as California or Colorado, solar exposure was a valuable commodity. Agents and land sellers in the Berkshires did not yet have this awareness.

I made a few phone calls and double-checked some of my data. I visited the land, and all went well there also. Early in the afternoon I stopped by Mary Jane's office to try to make a deal. I would love to tell you that I was a great horse trader and that I got a great bargain. Unfortunately, I wanted the land and all the problems that went with it. I wanted very badly to move from my sister's old house. I put in an offer for $10,000. Beyond the usual terms of sale, I added that the sale was contingent on the sale of my father's old trailer which was scheduled for a few weeks in the future. Without the money from the sale of the trailer, there could be no land purchase.

Mary Jane was a sharp real estate agent. Before she even mailed the signed offer to the owner, she reached him at home on Monday night. He told her he would bargain and sell the land for $11,000. She called me, and I agreed. The next day I signed an updated offer sheet that she mailed to the owner. The deal was done. I had paid too much money for a piece of land with no power in the outback. But I had purchased a piece of land that had perfect solar exposure, was a good wind site, and had only one set of neighbors, who were just the right kind. From my days of buying musical instruments, I knew that paying a little too much for the special guitar for my own personal use was never a bad deal. I applied this principle to my land purchase.

Everything had gone too smoothly up to this point. Toward the end of the week, Mary Jane got a call from the seller. The monkey wrench had fallen into the gears. The seller's wife, the co-owner of the property, would not co-sign the offer sheet. She wanted $12,500, period. After several days, I agreed to pay the full price. Again we sent a signed offer sheet to the sellers. Four days later Mary Jane got a call. This time the man was apologetic. He was not getting along with his wife, and she was using this land sale as a power play. She did not want the land, but she would not let him sell it even at the full price. I realized that I was back to square one. I went home and quit the project.

During the week or two when the offer sheets were going back and forth and it seemed sure the land was to be mine, I had complimented Mary Jane on her dealings with me. I told her I would be finishing the still unfinished parts of my sister's house over the next few months and that I would like her to try and sell it for me. As I have said before, Mary Jane was a keen real estate agent. She called me the next day to say she had someone who would be interested in the house.

Wind exposure for a potential wind machine.

I told Mary Jane I was not ready to sell the house, because it was not yet finished. What I did not tell her was that I needed a place to live. She suggested we show the house on Saturday, and that I should be present to point out the work that I intended to finish. I thought she was crazy, but I gave in and said yes.

We needed to set a price. I asked Mary Jane for her opinion. Her price was about 20% higher than I had expected. The couple looked at my house on Saturday. By Wednesday I had bargained and thrown in the tank of oil, wood stove, and kitchen appliances. The house was sold, contingent upon my purchase of the land.

When I was sure that the land deal had fallen through, Mary Jane decided that I needed the land and that she was going to make the deal happen anyway. Of course she also had her own interests: the land sale, the house sale, and her commission. She called the sellers and reduced her commission to virtually nothing. I would still pay $12,500, but the sellers would get much more money as a result of the reduced commission. Even the unhappy wife was happy. I had paid a few thousand dollars too much for the land, but I had sold my house for $8,000 more than I had expected.

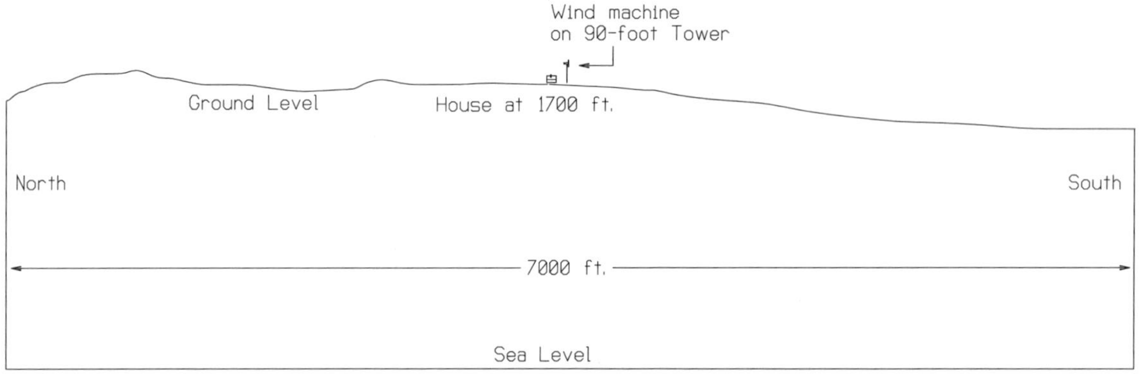

North-South Wind Exposure

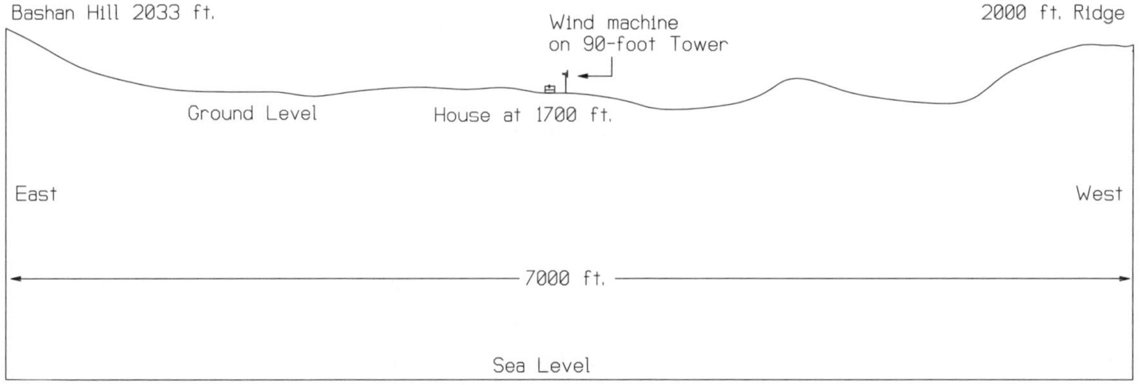

East-West Wind Exposure

IT WAS NOW about the beginning of April. My friend Richard was truly amazed that I had called him about living without electricity one Saturday, and then in about three weeks, I had agreements for the sale of the trailer, sale of the house, and the purchase of the land. I was less amazed about all the events that had just occurred and more worried about the rapidly approaching day of May 18, 1981. On that day the house would be sold simultaneously to the closing on the land. Several weeks before this date I would be selling the trailer.

I had also bought a 1951 Willys Jeep and plow. This was to be used to plow the driveway and serve as a cheap tractor-type vehicle while I cleared the house site. I had paid $600 for this jeep. I did not inspect it as thoroughly as I should have. It needed a lot of work just to get it ready for the trip across town. In the next six weeks I would have to finish the house, get the trailer ready for its sale, find someplace to store my things, and move to a tent on my new land.

CHAPTER 4

The Move to Bashan Hill Road

BY MAY 18, I had completed the work on my old house, transferred the properties, and moved to the new land. Suffice it to say that I had done too much in too little time and that I was exhausted. After the closing at the lawyer's office in Northampton, my little dog Rolo and I headed to our new home.

I had found a circus-tent-shaped, old army tent at my sister's house, left behind by my ex-brother-in-law and his Boy Scouts. I had cut off the top of a tree on my new land and had used the ten-foot high stump as a center post for the tent. This was stuffed with tools and belongings that could survive summer in a tent. Next to the tent was a giant pile of leftover lumber and used building supplies.

Over in a little abandoned orchard, near the intersection of two stone walls, stood a nine-foot square dome tent. This tent wasn't very good when I bought it at a discount store ten years earlier. I had used it over the years whenever I was in between places to live. I had upgraded the tent for this project by covering the moth holes with duct tape. Over the external frame I attached a polyethylene fly that would keep most of the rain from ever touching the old canvas. Inside the tent, I had a footlocker for my clothes, and this doubled as a table for a kerosene lamp. I would sleep in an old sleeping bag on an old foam mattress.

Outside, next to the tent, I installed a two-burner, schoolhouse wood stove with a damper and two lengths of stovepipe straight up into the air. Each day, 100 feet away, I dug a fresh hole to bury the contents of the white joint compound bucket that served as my makeshift toilet. Fifty feet away from the tent, nestled under a stand of 25 foot Douglas firs, I had placed my simple office, an old metal desk that contained documents and paperwork. I covered the desk with a sheet of polyethylene. I kept my valuable guitar locked in the hatchback of my car. I attached a rope to an apple tree so I would be able to tie up Rolo when trucks or other heavy equipment came to the house site. The rest of the time, he slept in the tent and roamed his new retirement home. Most mornings he trotted down to the Cooks' to mark the trees for Annie, the black lab.

The first night in the tent was just plain cold. I started work the next morning. Within a few hours I was confronted with an event that made me question my whole move. Sometimes there are obstacles that cannot be foreseen.

Rolo had been running free. After all, this was the first day at his new home. In the distance his barking changed to cries of pain. He came running back to me with a nose and lips full of porcupine quills. We rushed to the vet to have them removed. Luckily, Rolo was not one of those aggressive dogs who attack a porcupine and get a mouth full of quills. The vet pulled the quills without too much effort. I learned from the vet that the quicker quills are pulled the less they penetrate. If Rolo found another porcupine, I should pull the quills myself rather than waste the time to drive 45 minutes to the vet.

My little dog Rolo.

I drove home with a drugged little dog sleeping on the seat beside me. I was devastated. I thought I had solved all the problems of my old home only to now feel that while Rolo could run free with no fear of cars, he could not run free because of porcupines. I tied Rolo up for several days and walked him on a leash around the boundaries of my property to let him mark out his territory. I hoped to scare away the porcupines. A few days later when I let Rolo run free again, he found another porcupine. This time he had only a few quills in his nose. The first time he learned the porcupine could hurt him. The second time he only tried to smell this strange and dangerous critter. Rolo only weighed 17 pounds. In a few minutes I had rolled him up in a blanket with only his head free. I easily pulled out the freshly inflicted quills.

The porcupines never moved away because of me or my dog. That was the last time Rolo and I dealt with the quills because Rolo had adapted to his new environment and his new neighbors. I too adapted. I had bought this land and could change it with my machines, but it was not really mine.

We had a frost each night for the next week. At the end of each work day, I drove to the little river one mile away on Parish Road to bathe. I stood in the icy water up to my calves and soaped up. This was in itself a cold way to bathe, but the worst part was rinsing the soap off. If I could have bathed in a lake, I could have quickly plunged to rinse off. Instead I had to stand in the icy water and douse myself with many containers of icy stream water. For privacy I hid under the green bridge that spanned the river.

For the first few weeks I had three major objectives: clear the land for the cellar hole, plant a garden, and design the final house plan by the light of a kerosene lamp in the tent.

I wish I had not had a garden that first year: it was too much work at a time when hours were short. I had had my first real garden the previous year at my old house and loved it so much that my reasoning was not clear. I knew I would not have a refrigerator. I would count on the fresh food that would not spoil while it grew in the garden. But I ended up working until dark after long work days on the house. Garden work was one extra chore I did not need.

Before I sold my house, I made a deal with my neighbor, Gerry Landry. I traded him a cabin I had on my land, which we put on skids and dragged up to his land. In exchange for this, he was to come across town with his tractor and make a garden for me. At that time Gerry worked full time, and after hours built his house. Most of his house was built with traded labor. Unfortunately, Gerry was always so overcommitted that it was hard to get him to pay the help back. I nagged him for a week. He finally came, but his tractor and plow did nothing more than a terrible job turning the soil. He couldn't bring his harrow across town, so the sod he did turn over was left in giant clumps. In a half of an hour he was gone. I was not much closer to a garden.

A few days later I got Paul Sena to come over with his tractor and PTO rototiller. This commercial rototiller was great except that the area that I had chosen had many boulders a few inches below the surface. Paul tried his best. He broke several tines on his tiller on the boulders and quit. Amazingly, he refused any pay at all. I turned over the rest of the garden with a shovel. That first summer I had a small garden choked with weeds, yet it produced all the vegetables I needed.

There was no water for the garden. The well had not yet been dug. Furthermore, even when the well would be dug, there would only be water if I ran the generator to make electricity for the pump motor. The summer of the house project was dry. This was great for the house building schedule, but tough on the garden. One evening, I decided to get water for the garden with my old jeep. I put four six-gallon white buckets in the back and drove one mile to the green bridge. I filled the buckets and carefully placed them in the rear so that they would not tip over. I wish I had a video of my ride back to the garden.

This vehicle was in very poor condition. As is true of most of these jeeps, New England's winter road salt had rotted away the body. In the rear, the body mounts that should have attached the body to the frame were long gone. The rear was mounted strictly by gravity. Every time I drove over a bump, the buckets slopped out a little water. I drove slower and slower. I weaved backed and forth on the road trying to find the smoothest trail.

The green bridge one
mile away on Parish
Road.

When I got back to the garden, I had four half-full buckets of water.

After the first week of nightly frosts, we received three days of 90° F. weather. These three days happened to be the same days I had chosen to clear the land for the cellar hole. This area was covered with 25 foot tall, overgrown Christmas trees. I had hoped to have my nephew working by then, but he was still involved with his upcoming graduation from high school.

This was perhaps the most miserable job of the whole project. I was in good shape, but I was unprepared for this heavy work at the same time the weather was so unseasonable. I cut down the trees and limbed them with a chain saw. Running a chain saw and cutting wood in 90° F. heat is no fun. Then I chained the trunks to the jeep and dragged them off. I spent the whole three days wrestling with that miserable old jeep. My only rest came when I would blow a brake line or a hose on the jeep and have to stop to fix it.

I HIRED ZACK DONOVAN to dig a cellar hole for me. I liked Zack. He was the father of six kids. I had gone through grammar school and high school with three of his sons and daughters. He liked me, too. He always referred to me as a good kid, and a good ballplayer (he had been my Little League coach). Zack had a small 'dozer with a backhoe on the back. His skill was exceptional. He seemed to have his hands and fingers surgically attached to the levers of that machine. All he needed to do was visualize what the machine should do, and the 'dozer responded. Under most conditions, he could do more work with that small 'dozer than most people could do with a much larger machine.

The only unpleasant part of the job was break time. Several times the conversation would turn to my father. Zack had not liked my father. Some days I just did not want to hear it.

Zack's dislike for my father stemmed from an incident in 1952. Zack had not been able to buy our house and 40 acres that my father had for sale. Zack and his family lived in Chesterfield, the next town. Zack's wife had stopped to look at the house. She said they would be interested in buying it, but that her husband was on a hunting trip for several weeks. God only knows if Zack's wife thought my father should have put the house sale on hold, or if my father promised too much in his eagerness to sell. Alberie Albert of Albert Farms was a great character and my father's close friend. He showed up one evening to announce that he was buying our house and land. He carried a shoe box under his arm from which he produced cash for the full purchase price of the house. The deal was done before my father had even met Zack Donovan. To further irritate the bad feelings, Alberie put 37 acres into potato fields, and later he sold the house and only three acres to Zack, who was related to him by marriage.

Zack charged me a low price for digging the cellar hole. His price per hour was very fair, but he was so efficient, the number of hours was small. I was the contractor on this project. I made a number of errors on the project that a full-time, experienced contractor probably would not have made. I will try to relate some of these mistakes to you, not because I want to complain, but because you may find yourself in a similar situation someday, and you may be better equipped to deal with it. One misjudgment concerned the depth of the cellar hole.

"The digging is getting awful hard," Zack announced toward the end of digging of the cellar hole. He explained that I only really needed a seven-foot cellar and how the foundation should stick up a little higher. I should have asked about bringing in a bigger 'dozer. Zack probably would have seen that I could have afforded to complete the job. As it turned out later, I spent $500 on fill and several hundred dollars more on grading to correct the problem of a foundation that was too high out of the ground. I think the real problem with being your own contractor is that the project is your own house. The workmen come up to you and recommend a quick solution to a problem. You take their advice quickly. You have too many things to do. You can never say, "I will think about it and call the owner tonight."

A second mistake set me on a more careful tack for the remainder of the project. I had money to spend on this project. I had made some profit on my sister's house. I had improved the poor inherited trailer and turned it into cash. I also had some savings. After the purchase of the land, I had about $25,000 for a septic system, a well, materials for the house, and groceries. This was, of course, very little. But for me, it was a bankroll, and I wanted to enjoy handing out some work to the people with whom I had grown up.

Alan, one of my distant relatives by marriage, was out of work. He operated heavy equipment on construction sites. I wanted to be a do-gooder. I asked Alan if he wanted to do my septic system, well pipe trench, and grading with his own backhoe. He jumped at the chance. I think all would have gone well if Alan had not been called back to work.

The first job I had for Alan's backhoe was to dig a 200 foot, four-foot deep ditch from the well to the house foundation. Alan drove the backhoe over one evening by the back roads from the next town. By this time, Alan had gone back to work. He proposed a cheaper price per hour to have his seventeen-year-old son Justin run the hoe. This seemed fine to me. I could help Justin, too, with my bankroll.

All did not go well. It took Justin eight hours to dig the ditch. I got a bill for eight hours at $25 per hour. This was $200. Zack Donovan had charged me $25 per hour and less than $400 to dig the whole cellar hole. I made a few calls to contractors after I received the bill. All questioned said the eight hours was absurd. A month later, this became even clearer when I had a similar ditch dug by Zack in two and a half hours.

I was in trouble. I had made a deal with part of my extended family and it was going sour. I called Alan and questioned him about the amount. He told me he got $30 an hour and he took $5 an hour off for Justin's lack of experience. He said it was hard digging. I talked to Alan's brother and sister-in-law, who basically told me Alan could be a little stubborn to talk to and they did not want to get involved. I made up my mind. I wrote a check the next day for the full amount. I thanked him, but politely suggested that the digging was too hard for his backhoe. I would hire a bigger hoe for the rest of the work.

Part of the problem here was that Alan had gone back to work but still wanted to follow through with his commitment. Justin was a great kid. He tried his best, but he was not worth $25 an hour. The biggest part of the problem was the backhoe. It was old, and the engine needed to be rebuilt. Justin did a good job of running the levers; the engine was just way too tired. The $100 extra I had spent on the ditch was well spent. It put me back on earth. This was a house-building project for me, and work for the people I hired. I could not be the benevolent big spender.

I HAD DESIGNED MY NEW HOUSE before I had moved to the new land. Many of the decisions concerning the new house were in direct consideration of my plan to live in a house that would be powered by alternative energy. The house had to be a passive solar house. There could be no fans or circulating pumps to move and store solar-heated water or air. The future alternative electrical system would not be able to meet the electrical demands of the electric motors. Likewise, there could be no backup propane furnace. Conventional furnaces require heavy electrical loads for blowers and circulating fans or circulating pumps.

This house would have to be heated by passive solar and wood heat. There would have to be enough insulation and thermal mass such that the house could maintain a temperature above freezing if I were to be away for days, or even weeks. I did not want to be one of those country folks who could not leave home in the winter because the fires needed to be tended. Passive solar heating matched with passive heat storage and great insulation was the only way to have a house that could be left with no backup heat and not freeze. During the evenings in my tent I finalized my house design. I plotted out on graph paper the final dimensions and roof pitches. I subdivided the two floors into rooms. Basically, I took my house design and made it buildable in terms of engineering and construction. After I made the decisions of spans and loads and bearing walls, I got all the dimensions down to the precise measure and drew up an exact house plan. The house plan still needed many final decisions that were later made the night before I began the actual construction.

CHAPTER 5

Passive Solar House Design

THE SUN IS A SOURCE of both light and heat energy for our planet earth, passing as rays through space and through the atmosphere that encircles the earth. Different wavelengths of these rays are the visible colors and radiant heat. Of course, there are a few more components of sunlight, but we will mostly concern ourselves with the heat that comes from the sun.

When the sun comes up in the morning it changes the world from a colorless dark to a color filled light. The colors of objects are determined by the parts of the light from the sun that are absorbed and the parts that are reflected. We all have noticed that a very dark object in direct sunlight gets hot. This is because the dark object absorbs more of the energy from the sun. When the sun comes up, the solar energy passes through the atmosphere to be absorbed by the surface of the earth. The earth is warmed.

A special condition exists on the earth. Some of the heat that is absorbed on the surface stays to maintain the stable temperature of the earth to which we are accustomed. Other planets behave differently. The heat warms the surface when the sun shines, only to be quickly dissipated back out to space when the sun stops shining. The daily temperatures of these planets fluctuate greatly.

The solar energy that passes through space and through the atmosphere of the earth is in a short wavelength form. These waves are absorbed by the surface of the earth, thus raising its temperature. At night, the surface that has been heated radiates long wavelength infrared radiation out toward space. This long wavelength radiation

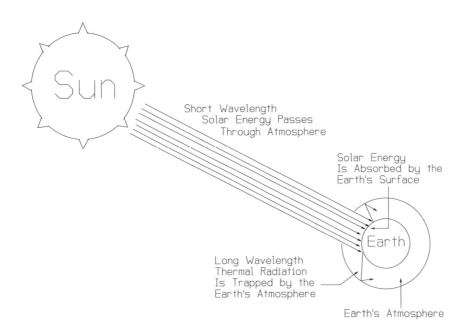

Short Wavelength Solar Energy Passes Through Atmosphere

Solar Energy Is Absorbed by the Earth's Surface

Long Wavelength Thermal Radiation Is Trapped by the Earth's Atmosphere

Earth's Atmosphere

Solar energy passes through the earth's atmosphere and is retained.

does not pass through the earth's atmosphere as well as the short wavelength solar energy that entered the atmosphere from the sun. Much of the long wavelength radiation returns to earth. In principle, the earth and its atmosphere let the heat from the sun in easily, but do not let it out easily.

The earth is habitable to mankind because of its ability to hold the heat from the sun through the night. As I have said before, this creates a stable range of nighttime and daytime temperature fluctuation. In the summer the difference between daytime and nighttime temperatures is in a comfortable range for us. However, in the winter in New England where we live, we must heat our home so there is less of a fluctuation between the daytime and the nighttime temperatures. We also need to heat the house to be warmer than even the warmest outside temperature of the winter day.

A solar-heated home simulates the sun's heating of the earth, and the earth's storing of that heat. The sun's solar energy enters the house and is absorbed by the inside surfaces. Ideally, as much of this heat as possible is retained to keep the house warm until the next day when the sun shines.

If a solar heated house could be heated in a day by sunlight and be insulated perfectly, the temperature would stay exactly the same and not fall at all during the night. In fact, if such a house were perfectly insulated it, theoretically, would never lose any heat and would never even need to be heated again. Of course, once you opened and shut the door to go outside, you would cause a loss of heat, and the house would need more heat to balance the loss.

Solar heated houses are usually grouped into two categories: active solar houses and passive solar houses. In the day, active solar houses use fans or

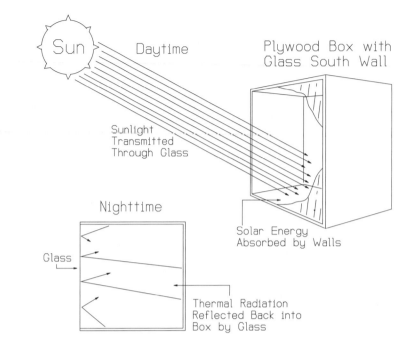

Daytime

Sun

Plywood Box with
Glass South Wall

Sunlight
Transmitted
Through Glass

Solar Energy
Absorbed by Walls

Nighttime

Glass

Thermal Radiation
Reflected Back into
Box by Glass

**Solar energy passes
through the south facing
glass of our solar box,
but thermal radiation
does not pass back out.**

pumps to circulate the incoming heat through masses of material to store the heat. At night, the fans or pumps circulate the heat back into the living space to do a better job of heating the house. The electrical energy that is needed to power pumps and fans in an active solar house is a significant and consistent load. The home I was designing was going to be powered by alternative energy. The greatest factor in the success of an alternative energy system is conservation of energy. I chose to avoid the loads of fans and pumps and use passive solar heat.

Passive solar heat is passive in the sense that it does not utilize the moving parts of the fans or pumps of an active solar house. A passive solar house functions similarly to the earth and its atmosphere. It lets the sun shine in to heat the house and retains a portion of that heat to keep the house warm until the next day when the sun again heats the house.

The simplest example of a passive solar home is a box, let us say with dimensions of eight feet by eight feet by eight feet. The cube is oriented such that one vertical side faces exactly true south in New England. This south face of the cube is clear glass. The other five sides (the other three walls, the ceiling, and the floor) are made of an opaque material. For our example, we will choose plywood.

The sun in the northern hemisphere rises in the east, passes through the southern sky, and sets in the west. For most of the sunny hours of the day the sun will shine into our passive solar house box. The house will be heated as the sunlight is absorbed by the plywood interior walls. At night the heated plywood will emit long wavelength radiant heat that does not pass through the glass window as nicely as the short wavelength solar energy that entered from the sun.

Sunlight passes through the glass into the passive solar box and is absorbed by the walls and floor. The glass does not allow thermal radiation to pass through it. The main problem with this passive solar house box is that it will not stay warm enough at night when the sun is not shining. The first part of the problem is that neither plywood nor glass is a good thermal insulator. The heat inside the box will quickly pass through the plywood and glass to the outdoors. We need to insulate the box. We will cover the exterior of the five opaque sides of the box with eight inches of fiberglass insulation. We will upgrade the south facing glass wall with double-pane or insulated glass.

The second part of the problem of lack of nighttime warmth in our passive solar house box is the box's inability to store the heat that comes into it. Wood is a material that does not possess the proper characteristics needed to store heat or thermal energy. Concrete or stone has a good ability to store thermal energy. If we placed a large black boulder in the box, the sun's energy would heat the boulder surface. The surface would gradually conduct heat to the interior of the rock. During the night the interior of the rock would conduct heat to the surface of the rock, which would in turn radiate the heat to the inside of the box. Since we have insulated our box, a lot of

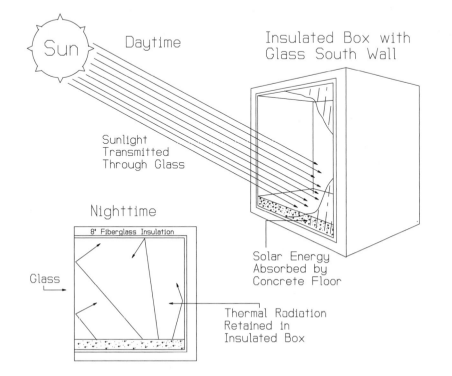

Solar energy heats the concrete floor of the passive solar box during the day. The insulated box retains heat at night.

the heat given off from the boulder would stay in our passive solar house box to keep us warm. A smaller boulder would store less heat; a larger boulder would store more heat.

I could never envision myself living in an eight-by-eight-by-eight-foot box. Furthermore, I would not like to give most of that space up to a big black boulder. The solution is to replace the thermal mass of the boulder with the mass of a concrete floor in the box. Now the sun will shine into our box and store heat energy in the floor. At night the floor will give off the heat to heat the box.

It is pretty easy to see the need for the thermal mass to provide heat to the passive solar house at night when the sun does not shine. I like to view the function of the thermal mass from a slightly different perspective. The thermal mass is a heat sink, or reservoir of heat. When the house is cold, the heat from the thermal mass heats the house. Equally important, the thermal mass absorbs heat when the solar energy is plentiful. This action keeps the solar energy from overheating the house.

Over the years, I have seen many almost passive solar houses. They had plenty of south facing glass, but no thermal mass on which the sun could shine. Instead, the sun heated the standard floor that was not capable of storing the heat for the nighttime. The floor absorbed the sun's heat only to heat the air above it. The result was a 90° F. room or house on a sunny day in January. The windows had to be opened to let the heat out. Of course, the heat was long gone when it was needed in the evening.

The black boulder or the concrete floor is called a thermal mass. A passive solar house has two types of thermal mass. One type is direct-gain thermal mass. This is the example of the floor or the boulder. The sunlight directly shines on and heats the mass to store thermal energy. The second type is indirect-gain thermal mass. This is thermal mass in our passive solar house that does not get heated by sunlight directly shining on its surface. In the example of our box, we could make the walls and ceiling of concrete. These masses would be heated indirectly from the air in the house.

Our simple passive solar house box has glass or windows only on the south side. This is the ideal. The glazing on the south lets the sun in, but it is really as poorly insulated a wall as the plywood walls were before they were covered with fiberglass. The windows lose a lot of heat all the time. When the sun is shining through the window, the amount of heat lost is much smaller than the amount of heat gained. Optimally, we would like to cover this poorly insulated wall with insulation whenever the sun is not shining. It is rarely feasible to employ movable insulation in a regular home. I chose in my home design to delay any decision to use movable window insulation until after the house was finished.

The south facing glass makes sense in our passive solar house box because the thermal energy that the house gains during the sunny hours is much greater than the heat lost over the full twenty-four hours. Many people want to add a lot of glass to the east wall to get the strong morning sun and to the west to get the strong afternoon sun. This does not work because the amount of heat coming in for one or two hours in the morning or the afternoon is much less than the heat lost from the same windows through the entire twenty-four hour period.

As we mentioned above, the sun rises in the east, passes through the southern sky, and sets in the west. The layman often forgets about the summer to winter change of the sun's position in the sky. In the Berkshires, the summer noonday sun is approximately 70° above the horizon. In the winter, the noonday sun is only about 25° above the horizon. In our passive solar house box we have vertical south facing glazing. In the winter, when the weather is the coldest, the sun is low in the sky, and it is providing the most direct sunlight into this vertical glazing at an angle nearly perpendicular to the window. In the summer when the weather is warm, the sun is high in the sky. The sun barely shines into the vertical window. Consequently, we do not have to worry about the sun heating our passive solar house when the heat is not wanted.

In the Berkshires, vertical glazing allows maximum winter sun and minimum summer sun. It is difficult in a very cold climate such as New England to heat a house solely by passive solar energy. The house has to be superinsulated, which usually means foam insulation. When I designed my house, I felt this was questionable in terms of potential indoor air pollution. Any plastic, especially if it is not hard like glass, has the potential to release chemical gas as it ages. I felt some of this gas would always find its way into the house. Furthermore, a solely passive solar heated house would need to be incredibly tight to resist any infiltration of the outside cold air. I felt that a house this tight would cause a buildup of bad air from gas stoves, wood stoves, glues, and formaldehyde in building materials.

Another problem with a solely passive solar house in New England is the frequency of sunny days versus the frequency of cloudy days. It is not too difficult to design a house that can be re-heated the next day by sunlight. It is very hard to design a house that can sit and stay warm for several days or weeks when the sun does not shine at all. For this scenario, a house must have a truly gigantic thermal mass to balance the heat loss over many days.

I had spent a lot of time designing an earth-bermed passive solar house in the year preceding my project at Bashan Hill. This type of partially underground house could have the large amount of thermal mass to get through cloudy spells. The walls and floor of such a structure are in contact with the underground soil instead of the colder winter air. The heat loss is, of course, much less. I just did not want to live underground and worry about water infiltration. I did incorporate some of the principles of an underground house by employing in my passive solar house a cellar that was insulated on

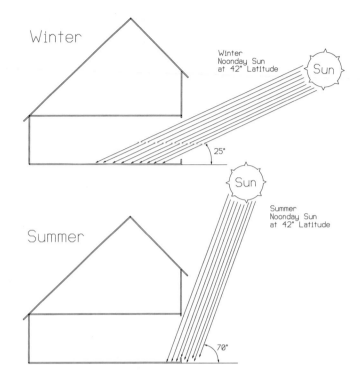

Winter

Winter
Noonday Sun
at 42° Latitude

Sun

25°

Sun

Summer
Noonday Sun
at 42° Latitude

Summer

70°

A large amount of solar radiation passes through vertical glazing in the winter, and a minor amount passes through in the summer.

the outside. This cellar is constantly heated in the winter by the 50° F. earth underneath the floor. This helps keep my passive solar house from freezing when it is left for long periods of time.

I chose to design a partially passive solar heated house. I wanted a reasonable amount of air infiltration. I planned to supply the balance of the heat by wood heat. On sunny winter days the sun would heat the house. At a certain point in the early evening, the temperature would fall and the wood stove would be lighted. Only a small wood stove would be needed because the stored heat in the thermal mass would be providing the balance of the heat for the house. On days when the sun would not shine, the wood stove would run all day and night.

The ideal partner for a partially passive solar house is a small thermostatically controlled furnace. In the evening, when the temperature begins to drop, the furnace provides just the right amount of supplementary heat to keep the house at an even temperature. I chose to heat with wood because it was local, do-it-myself, and cheap. A big advantage of heating with a wood stove was that it did not need any electricity as a small furnace would. The only disadvantage of the wood heat would be that the stove would need to be constantly, and not automatically, filled with wood. Which meant I would have to be home for the supplementary heat to be available. I mentioned that it is difficult to design a solely passive solar house for a cold climate like New England. My goal was to design a passive solar house that would not drop to freezing temperatures when left with no supplementary heat. The water would not freeze, and the plumbing would not burst. In New England, the subzero cold snaps are luckily coupled with clear sunny days. Even in the coldest snaps when the house would be left for weeks, the extremely high amount of solar heat on a clear day would balance the extremely cold subzero night. Because the house would be so well insulated for the passive solar heat, it would be easy to bring the house back up to normal temperature upon returning and starting the wood stove.

I had learned the principles of insulation, construction, and passive solar design in order to design a house. I now would need to make a final decision on a house size and type and translate my theoretical knowledge into a finished set of house plans.

CHAPTER 6

The House Design

DURING THE YEAR that preceded my move to Bashan Hill, I considered many different passive solar house designs. I really liked three designs that were so different they could not be melded into a hybrid. I liked the heat-holding capabilities of an earth-bermed or underground house for the very cold climate of 1,700 feet above sea level in the Berkshires. I also considered a modern sort of California-type house with clerestories. It was appealing to have a unique home. My third consideration was a New England saltbox house, which I chose as the decision deadlines approached.

I liked the idea of traditional lines in this backwoods New England setting. This house was to be only a piece of the bigger picture, a self-sufficient homestead. I could not see the house as a work of architectural art. Furthermore, I was not an architect. I had only a small amount of expertise and talent in house design. My strength and interest lay in designing a house that was non-toxic, made mostly of local materials, and efficient, especially with respect to the future alternative energy system that would power the house.

The saltbox is generally credited with being the earliest passive solar house. The first saltboxes were in England. They have been popular in this country since the English colonization. The front of the Colonial saltbox was oriented to the south. The majority of the windows were on this south side, which was two stories high. The north side of the house was only one story high with few or no windows. The solar heat could enter the house on sunny days through the large area of windows on the two stories on the south side.

A traditional saltbox.

The north side did not lose as much heat where the sun did not shine, because it had only one story of few or no windows. Some of the saltboxes had a woodshed storage area on the north side. The sloping roof continued downward at the same angle from the main house until it was only five feet off the ground at the outside of the storage area. This storage area provided more protection from the cold winds from the north.

The early saltboxes predated the coming of the wood stove in the early 1800s. They were heated by much less efficient fireplaces. During the coldest snaps of a New England winter, the fireplaces were not capable of heating the whole house. The interior rooms were broken up so that the inhabitants could shut off much of the house and retreat to a smaller area. I grew up in a 1768 colonial farmhouse in Worthington. In that house, the early inhabitants would retreat to the 42-by-16-foot kitchen and hearth when the −20° F. weather set in. My house would utilize many of the basic principles and the basic dimensions of a classic saltbox. The English saltbox often had a steeper roof pitch than the New England counterpart. My roof would be a steeper 12-12 pitch, or 45° angle, on the larger northern roof. Also the southern roof would have nearly as steep a pitch. This English design would give more living space upstairs. This would also enable the upstairs closets to be in the north walls of the rooms upstairs. The closet wall in common with the bedroom would be full height. The north wall in the back of the closet

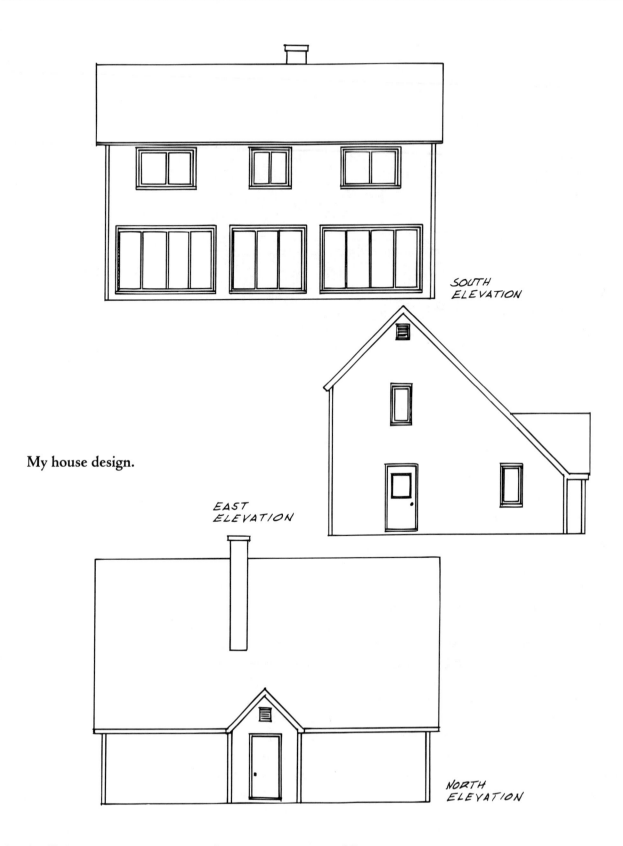

SOUTH
ELEVATION

My house design.

EAST
ELEVATION

NORTH
ELEVATION

would be short, because the ceiling would be the 45° sloping roof. Since the closets would run the length of the north wall of the upstairs, they would function as a dead, cold-air space, insulating area, much the same as the north side woodshed storage area had served on the early saltbox.

There would be no windows on the north wall of the first story. This would be a continuous, well insulated wall. The east and the west walls would have only one medium-sized window per room. Again, there would be the minimum amount of heat-losing window area on these non-south sides. The first floor south side would be all windows to harvest the maximum amount of solar heat available. The second story of the south side would be about half windows. The reduced number of second floor windows was necessary because there would be no thermal mass to absorb the heat from the sunlight. I did not want the upstairs to overheat.

In the last chapter, I discussed the importance of thermal mass in a passive solar house. A passive solar house must have the correct amount of heat storage. This must be sized in relation to area of the windows on the south side. The thermal mass itself must have the correct volume for the specific heat rating of the material of which the thermal mass is composed. Furthermore, the volume of the thermal mass must have a specific ratio to the amount of area that is heated either directly or indirectly by the sun. If this sounds complicated, that is because it is. For me, in 1981, it was even more difficult. There were conflicting ideas about this new technology. I made design choices based on careful calculations and educated guesses. I will describe only the final design, not the specific decisions leading up to it.

The thermal mass or heat storage for this house would be provided by a cement slab covered with Vermont slate. This slab would be nine feet by thirty-six feet. The thirty-six foot direction would be from east to west on the south side of the first floor of the house, exactly in front of all the first-floor, south facing windows. In the winter, when the sun is low in the sky, the sun would come in at an angle such that it would shine on the full nine foot width of the slab. In addition to the floor slab, there would be three, four-foot-high-by-nine-foot long mortared stone walls, perpendicular to the south facing glass. These stone walls would receive direct sunlight from the south side windows on their east sides when the sun was shining from the east and on their west sides when the sun was shining from the west.

Most first-floor slabs are poured on the ground with no cellar beneath them. I wanted a full cellar in this house. I have always found a cellar to be very important in a house. I would definitely need this area for my homestead. I chose to have a slab that was suspended above the cellar. This would be more expensive and would involve more labor. To me, it was worth it.

A suspended slab is much more difficult to build than a slab on grade because concrete is structurally weak. Concrete is a hard material, but it does not have the tensile strength of a material such as wood or steel. If you took

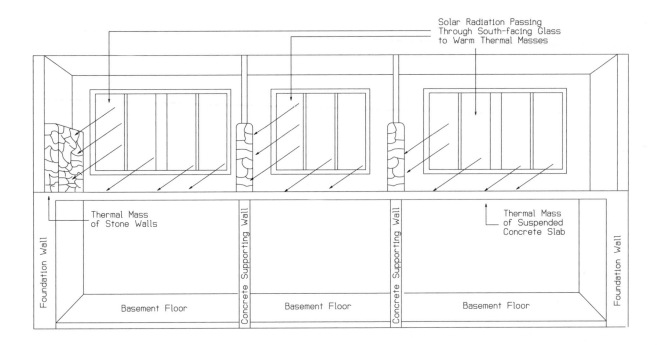

Solar Radiation Passing Through South-facing Glass to Warm Thermal Masses

Thermal Mass of Stone Walls

Thermal Mass of Suspended Concrete Slab

Foundation Wall

Concrete Supporting Wall

Concrete Supporting Wall

Foundation Wall

Basement Floor

Basement Floor

Basement Floor

Thermal mass to absorb solar energy passing through the south facing windows of my house.

a 2x10 plank and spanned a brook, the tensile strength of the wood would support you as you crossed the river. If you cast a piece of concrete in the same dimensions and attempted to walk across the river, the concrete plank would break much the same way a piece of glass would. The concrete needs to have something in it that does the equivalent structural job of the fiber in the wooden plank.

Concrete highway bridges add this "fiber" with a system of steel rods that are placed in the concrete when it is poured. The concrete hardens around the steel rods. For my suspended slab, I would need to add steel reinforcing rods.

Concrete is heavy. A typical house foundation has vertical forms, one on the inside of the wall and one on the outside of the wall. These forms keep the wet concrete in position until it hardens. They need to be only strong enough to hold the concrete in place because the weight of the concrete is supported by the footing and the ground at the bottom of the wall. I would need a one-sided horizontal form to support the bottom of the suspended slab, but this form would have to support the full weight of the slab.

The forms that support a suspended slab on a large job use either three-quarter-inch plywood or metal for the floor of the concrete. The forms are supported by metal posts whose bottoms sit on the cellar floor below. I would not have this equipment. I would be able to approximate the support of the metal posts by using 2x4s every two feet, if I heavily braced all of them together. If I let any of this form sag while the concrete was still hardening over the first week, the slab would develop cracks. I could not really afford the three-quarter-inch plywood. There would be no other place in the house where I would be able to re-use it. Furthermore, the plywood would have to be covered with oil to keep the concrete from sticking to it.

My final brainstorm was to substitute two layers of half-inch plywood for one layer of three-quarter-inch plywood. Under the concrete and on top of the plywood I would carefully place a continuous sheet of polyethylene to keep the concrete from sticking to the plywood forms. After the slab hardened, I would be able to remove the plywood and re-use it on the roof of the house.

When I first considered a suspended concrete slab, I asked my neighbor Gerry for some advice. He suggested that I talk to his friend Roger Gunn, who had had a business pouring concrete. Roger good naturedly offered me help. A few months before, Gerry's fiancee and I had sung for Roger's wedding. He liked me and really wanted to do me a favor in return. Roger explained the basics of the project, but was a little too imprecise for the engineer in me when he talked about reinforcing the concrete. He also seemed to imply that a twelve-inch slab would be stronger than a six-inch slab. He would be doing the job by the seat of his pants, not by structural engineering tables.

Later, when I was preparing to build the house, I discussed the suspended slab with my neighbor Bob Cook. Red flags went up everywhere. Bob had a lot of experience with bridge construction on his ironworker jobs. Bob was an excellent student of ironwork from an engineering and theoretical viewpoint. He lent me a few books on the structural design of concrete. He told me I had better have a exact design based on the weight (in this case the thickness) of the concrete and the span of the slab from one supporting wall in the cellar to the next supporting wall. The design would dictate how many reinforcing rods I would need and how thick each rod needed to be.

I spent the next rainy day at the library researching the design of suspended slabs. Roger's slab design may very well have never failed, because ultimately he may have overbuilt it by his seat-of-the-pants method. I wanted a design with no guesswork. If this slab ever failed, I would need to take my house apart to just get the pieces out. I was worried. I was in over my head.

The great part about growing up in a small town is that you know everyone. There is always someone to turn to who has the experience and expertise you need. Usually it is happily given, and often for free. Mike Lucey saved the day for me on this problem.

I talked to Mike at the general store about my house. When he heard about my problem, he offered to take my specifications to the engineering office where he worked to have the plans drawn up. I was out of trouble. I had plans for the suspended slab. Mike explained to me that a twelve-inch slab was no stronger than a six-inch slab. The twelve-inch slab would only weigh more and would require twice as much reinforcing rod to support it. This meant a twelve-inch slab would cost twice as much for materials, but would result in no benefit to my home. I changed the twelve-inch slab design I had settled on with Roger to Mike's six-inch slab design. At this point the foundation was already poured. As a result, I have a six-inch step-down to the slab from the wood floors in the downstairs rooms. In retrospect, this has become a design enhancement. All the slate-covered slab areas have become slightly separate sun spaces.

The basic house design was completed. The actual dimensions of the saltbox would be 36' by 26'. There would be 936 square feet on the first floor. The second floor of a saltbox has a full height wall on the south side of the second story, but the north wall is actually zero height because the roof slopes down to the top of the first floor. In this house there would be a full-height, seven-foot wall eighteen feet back (to the north) from the south wall. This would make a 36' by 18', or 648 square foot, second floor. All the closets would run down the north wall of the upstairs with the slanting roof of the saltbox inside of the closet. This would normally be wasted space in a saltbox. In this case the house would have 180 square feet of closets that were not taken out of the living space of the second story. Ultimately, this small compact house really would be the equivalent of a 1764 square-foot home.

THE FIRST THING I had to choose for the house, after I put the basic house design and the basic dimensions on paper, was the windows. The house would have to be built to a plan such that the windows would fit into their predescribed holes when they arrived, after the house had been framed and sheathed. I chose Marvin windows because, in 1981, they were the only all wood and glass (no plastic) windows I could find. These were dual-glazed windows. Window quilts for nighttime insulation would have to be designed long after the house was finished. Next, I designed the inside room layout with respect to the outside window dimensions.

One of the oddest things I did with the design of this house was to have a first-floor bathroom on the south side with a full wall of south facing windows. The bathroom is a neat design with an old claw-foot tub and old sink and copper pipes on a Vermont slate floor with stone walls. You can sit and take a bath in January and bask in warm sun. We would have no neighbors, so privacy did not seem to be an issue. However, we do now have mischievous children. It is strange to be seated in the pose of Rodin's *The Thinker*, to have your panoramic view invaded by a giggly, smirking child.

This bathroom design fit into the total function of my energy efficient house. I had taken heed of the design of the early New England saltbox with respect to its ability to close off rooms when the weather got extremely cold. I wanted a house that would, by the use of passive solar heat, maintain a temperature above freezing in winter if I were to be away. I didn't want to worry about the pipes ever freezing. I thought I could do this best by having the kitchen, bathroom, and all plumbing in the south side of the house. I could design a system

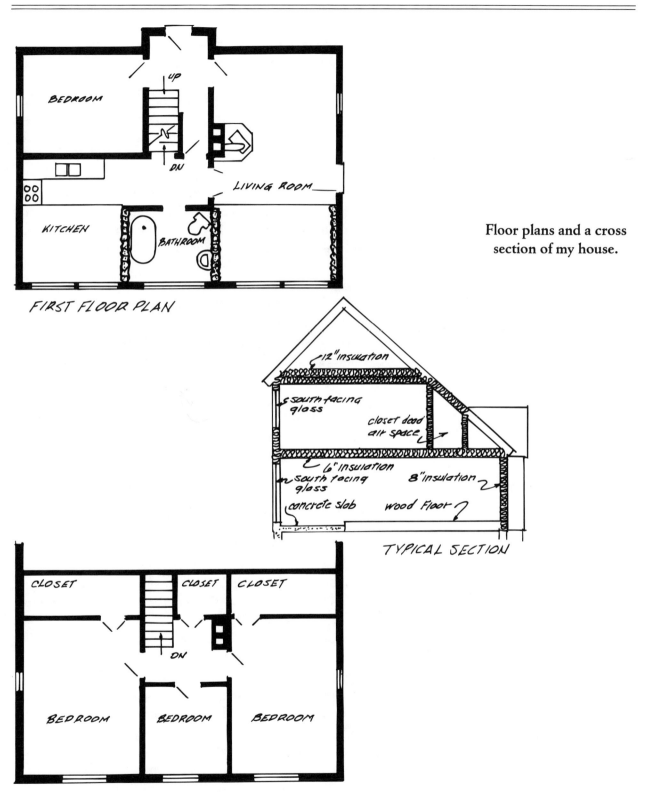

FIRST FLOOR PLAN

BEDROOM

UP

DN

LIVING ROOM

KITCHEN

BATHROOM

Floor plans and a cross section of my house.

TYPICAL SECTION

12" insulation

south facing glass

closet dead air space

6" insulation

south facing glass

8" insulation

concrete slab

wood floor

SECOND FLOOR PLAN

CLOSET   CLOSET   CLOSET

DN

BEDROOM   BEDROOM   BEDROOM

of doors and rooms such that doors could be shut when I left, and all of the south facing windows and thermal mass heat storage would be in a small area. This would intensify the passive solar heating ability of the house.

The central chimney and the wood stove would also be in this area. This would mean that the doors could be shut, as in a colonial saltbox, to minimize the area to be heated. Under normal conditions, the wood stove would heat the whole house. If I returned from a weekend and the inside was 45° F., then the wood stove could rapidly heat the smaller area up to 70° F.

The rest of the plumbing in the house would be no problem. The basement walls would be insulated on the outside with two inches of foam board insulation. This would probably mean the cellar would maintain a 40° to 45° F. temperature because the cellar floor would be constantly heated by the 50° F. earth below it. I had no problem using exterior foam insulation. Any off gassing would never enter the inside of the house or cellar.

As we discussed earlier, in passive solar heating a house needs to be very well insulated to retain the solar heat gained during the day. I was not willing to have foam insulation, though it has the best insulating capabilities. I chose to insulate this house with standard fiberglass. The wall studding would be different so that the fiberglass could be better utilized.

A standard house has walls of 2x4 or 2x6 studs with insulation in between the studs. Each wooden stud, positioned every 16", is a poor insulator compared to the fiberglass in between the studs. A conventional wall is like a quilted parka: there are puffs of down that keep you warm but there are also seams that do not. Super-warm parkas can be double quilted such that the seams of one quilted layer are covered by the down of the second quilted layer.

The walls in this house would have one wall of 2x4 studs with four inches of fiberglass in between. A second wall right behind the first would be of identical construction. However, the second wall would have its 2x4s out of phase with the first. Any 2x4 in the first wall would have the fiberglass part of the second wall behind it.

A great amount of heat loss in any house is the result of radiant heat passing through exterior walls. This happens a bit like heat radiating out a clear glass window. I chose to substitute the commonly used polyethylene vapor barrier that is placed over fiberglass insulation with two layers of aluminum-foil vapor barrier. The shiny aluminum vapor barrier would reflect the radiant heat that would normally pass through the wall. This was also another case where I could avoid the use of plastic and its potential off gassing into a tight home.

A horizontal cross section of an insulated double wall.

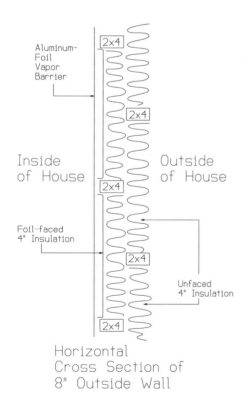

Aluminum-Foil Vapor Barrier

2x4

2x4

Inside of House

Outside of House

2x4

Foil-faced 4" Insulation

2x4

Unfaced 4" Insulation

2x4

Horizontal Cross Section of 8" Outside Wall

I planned to insulate between all of the interior walls. This would supplement the plan to shut doors to isolate the non-solar gain areas when the house would be left unattended. Also I planned to insulate the first floor ceiling to slow the natural passing of heat from the first floor to the second floor. I wanted a cooler upstairs for sleeping. I hated the thought of having too much heat from the wood stove on the first floor overheating my bedroom on the second floor.

In principle, the greatest heat loss in a house is upward. The hot air rises to the top of the room. Therefore the greatest amount of insulation needs to be in the upstairs ceiling. I planned to install twelve inches of fiberglass in the attic.

My years of study had been harvested to design my independent home in the woods. I could hardly wait to record and analyze the performance of my home during the first New England winter. First I would need to buy some materials and bang a few nails.

Alternating 2x4 studs in a double wall.

CHAPTER 7

Environmentalism and the Project

E VERY SUMMER, from the time I was six to the time I was nine, Larry Collins was my best friend all summer long. Larry was my best friend for the summers, very logically, because he was what was called a "summer kid," not to be confused with other categories like a "fresh air kid," or a New Yorker. Larry was a "summer kid" because his mother, Eleanor Collins, was a "summer person." That was the local vernacular for someone who had a home that he or she used only in the summer.

In the fifties, most of the small hilltowns like Worthington had a small year-round population and a much larger population during the summer months. Worthington had a population of 500 year-round residents. In the summer the population jumped to around 800. We had no lakes or resort attractions in town. Most of our summer residents were low-key people, seeking a quiet summer in the cooler hilltowns. Their summer houses did not have insulation or heating systems. When September came, the houses were closed up until the next summer. Over the years, most of these houses have been upgraded. Most of them are now primary residences. A smaller number are winterized, but are used only for the summers and week-ends for the rest of the year.

The Collins estate was very impressive. For any readers who are familiar with Worthington today, the old Collins estate is now the home of Ben and Frannie Albert. You may have noticed that I refer to "the Collins estate" and the "Albert home." The current Albert house is a large beautiful house with neighboring potato fields, a medical center, apartments for the aging, and four neighboring suburban type homes on small lots. None of these structures existed in the fifties. Worthington has grown.

The Collins estate was surrounded by flower gardens. Behind the house there was a beautiful barn. Unlike the large functional barn on my family's New England farm, this one was clapboarded and painted white. It had been built to suit the needs of a country estate: carriages, horses, a few chickens, and a few cows. The Collins' maid lived in the wing of the house that held the kitchen. The estate's guest cottage sat in a field a thousand feet up the road.

Larry Collins was actually named Larry Werner. His mother had been married to a man named Werner. Larry never mentioned any of this. He was happy to be called Larry Collins. My mother told me that while Larry was still a baby, his father had died in a car accident after he and Larry's mother had been divorced. As a grownup, I can now see this may have been the truth, or it may have been my mother's simplified story to her young son. To me and Larry, at that age, it was all the same. Larry had no father, as I had, so we didn't talk of it.

My father treated Larry as a part of the family. My father was the best father one could have until I was a teenager and his heavy drinking started to take its toll. He and my mother had both grown up with more did-not-haves than did-haves. Until the day they died, they loved to reach out and help the underprivileged friends of my sister and me. Larry had plenty of family money. He was poor in his lack of a father.

My mother and Eleanor Collins had general activities for those summers. On rainy days they played canasta with the Pease sisters, who also owned a summer house in Worthington. Helen, Mary, and Laura had outlived their husbands. They were older then and lived for many more years. My sister, Larry, and I always referred to them behind their backs by their town nickname, "the Hiccup Sisters." My mother and Eleanor played Canasta with the Pease sisters while they all sipped a few Tom Collins drinks (no relation).

On nice days my mother and Eleanor took us children swimming. For the first few years we swam in our pond, off Capen Road, in the lower fields of our farm. The 1955 hurricane and flood took out the dam in our pond. After that, we swam at the Chesterfield Bend, the Chesterfield Gorge, and Windsor Ponds.

Larry Collins was two years older than I was. He was also a good deal smarter. His young academic loves were biology and nature. We spent the smaller amount of the time at the swimming holes actually swimming and the larger amount of time exploring the surroundings. We were normally upstream catching fish in tin cans and pails, or off in the nearest swamp catching frogs. When we were at the Pease house we hunted for snakes, butterflies, bugs, and more frogs.

Bugs and frogs were an important commodity in our little world. These were the staples for the captives in our zoos. Larry had converted one room of his fancy barn for his menagerie. He had white mice, snakes, chameleons, fish, and a special South American tree frog that I once babysat for.

We set up a second menagerie at my house. We had a 32 foot by 24 foot ell attached to our large colonial farmhouse. Part of the ell had been the farm's unheated summer kitchen used for preserving food for the winter. My mother, who thought the world revolved around me, quickly agreed to give me an eight-foot by twelve-foot summer pantry for my menagerie. The shelves held the cages of snakes, mice, chameleons, and rabbit, and a meal-worm farm to feed the turtles and the snakes. My father donated a four-foot by six-foot cypress box, from the factory he managed, to hold the turtles. My mother would rarely venture into this room unless I coaxed her and held her hand. She was scared of snakes and hated mice so much she once avoided the living room for two days until I had recaptured the escaped white mouse that had set up living quarters in the sofa.

Environmentalism has been in my blood for a long time. As a youngster it was manifested in my love of nature. Even in the years Larry was not part of my life, I spent my time outside exploring every rock, ledge, brook, field, and tree of our 120 acres instead of watching Ranger Andy on TV after school. At college, I was a biology major. I was graduated in time to see the first Earth Day. I soon read *Silent Spring* and Adelle Davis's nutrition books. My great grandmother Alice White Fox was a Native American. At the age of 36, I earned a Masters Degree in Environmental Studies.

NEW ENGLAND YANKEES just cannot bear to throw anything away. My wife and I hate the clutter of too many things. We can't fit into the consumer world of "buy and throw." We do not buy much because we hate to own something that is not being used or not exactly what we want. When it comes time to clear things out, we have to find some person or organization to take the things and to use them.

When I designed this house, I did so as an environmentalist, a waste-not-want-not Yankee, a live-my-life-by-example sixties person, and a person who just did not want to be poisoned. These values were constantly influencing the design of this house.

I wanted my house to be more wood, glass, and stone and much less plastic, glue, and "man-made fibers." My first big choice was to frame the house with native lumber. In my locale, this is green lumber, unplaned and not kiln dried. Unfortunately, this lumber is often so fresh that it will still spit water when you sink a nail with a hammer. This wet lumber is one size when you cut it and nail it in place, and a smaller size a year later when it has dried. Boards become narrower, leaving cracks, and 2x4s become shorter.

Another problem with native lumber is the sizing. A sawmill is only marginally accurate. The framing lumber that you purchase at the lumberyard has been dried and planed. This planing actually corrects the imperfections so that each piece of wood has the same dimensions as the next. A rough cut 2x10 may be 10¼" on one end and 9¾" on the other end. These inaccuracies require both sorting of lumber and some experience of where to average out the discrepancies.

The biggest advantage of native lumber was the price. At the time I was buying materials for this house, rough cut lumber was 40 percent cheaper than dimension lumber. The second advantage was strength. A standard dimension 2x10 is actually 1½" by 9½" after it is sized and planed. The rough cut 2x10 is actually 2" by 10". This small increase in size actually makes the rough cut 2x10 30 percent stronger the standard dimension 2x10.

I chose native lumber for this house not only because of price and strength, but for the good of the environment. Local lumber does not have the environmental liabilities of the energy costs of heat for kiln drying and cross continental transportation. Old houses were often built with green lumber. The rule is to use all green lumber or all dry lumber. I chose to use all green lumber. Every piece of lumber would shrink proportionally. In a year, every wall would be an eighth of an inch shorter. The key was to hold off the installation of the insulation. This way the framing lumber could mostly air dry. The Sheetrock would not be installed for a year until the lumber had completely dried. This would keep the screws and nails from popping out from the surface of the Sheetrock as the lumber shrank.

When I was a teenager, Clarence Witter's mother bought Mrs. Arnold's house down the road from our farm. Later Clarence married Jane Nugent and lived in the same house with their young family. Across the street he ran his sawmill. Clarence was always interested in anything new, especially alternative energy. When I was building the house and thinking about a windmill, he wanted to get me a red pine log for a tower and install it with his cherry picker for free, just because he was interested. I happily ordered all my framing lumber from Clarence.

The choice of lumber for the wall, subfloor, and roof sheathing was a problem. Structurally, plywood is better than board sheathing. Plywood, when nailed in place, makes the equivalent of giant triangular bracing that keeps a house from racking or settling out of plumb. Houses built before plywood were built with corner braces instead. These were hard to incorporate into my insulation design. It was also twice as fast to install plywood as it was to cut and nail one-inch boards. For me the real decision concerned time. The environmentalist in me wanted to use native one-inch boards. Unfortunately, I had no place to live. I had to have this house closed in by September 15th. I knew I could not meet this deadline using board sheathing.

Environmentally, plywood is not as bad as it first appears. I was lucky. I was able to buy east-coast plywood. This meant a lesser energy cost in terms of transportation. Plywood has some environmental merits. The layers are actually made by turning a log on a giant lathe to produce a continuous sheet of ⅛ inch wood. These sheets are then glued together. All of the log is used. Furthermore, half-inch-thick plywood is structurally equivalent to one-inch-thick boards.

The bad part of plywood is the glue and the production required to make the wood and glue into sheets. Unfortunately for me, the real dangers of plywood did not become public until a year after I began building this house. The glue in the plywood is formaldehyde-based, meaning that it off-gasses formaldehyde fumes. Exterior grade plywood has a chemically stable, waterproof glue that has a very small amount of formaldehyde off-gassing. Fortunately, I had used exterior grade plywood even for the subflooring that is interior to the house. The truth about these glues arrived before I had completed the interior of my house. I then avoided all particle boards, interior grade plywoods, laminated furniture, and other formaldehyde based building products. The framing lumber would all shrink the same, but the plywood was dry and would not shrink at all. I felt that I had never been able to get a close fit on the plywood joints on any job I had ever worked. I guessed that the shrinking green lumber would only compress the spaces of the joints between the sheets of dry plywood.

I spent a lot of time thinking about lighting. Lighting would be one set of loads that would have to be met with electricity once the house was built and the alternative energy system was in place. The only real alternative to electric lighting was gas-light lighting. I had spent a few evening at Richard's house. Gaslights could never replace electric lighting for me. I didn't like the heat from them, nor did I like the faint smell. I certainly would not want fumes in a tightly constructed house.

I was so adamant about wanting electric lights that I refused to have any gaslights at all in my house. If I never had the mediocre alternative of gaslights I would be forced into action on an alternative energy system to power electric lights. The money I would save by not plumbing the house for gaslights would be money for my windmill. I planned to use plain old kerosene lamps until I finished the house and installed an alternative energy system. I found these lights only bearable for reading in bed, where I could hold the book right under the glass shade.

If you have never lived with kerosene lamps you cannot possibly imagine some of the hassles involved. A kerosene lamp does not give enough light to illuminate a room. It only illuminates the area right around the lamp. It sounds comical, but it is really frustrating to search all over the room for a paper or book, to later realize it was there, but the lack of light made it too hard to find. At other times I would come home after dark and not remember where I had left the flashlight. I would then feel my way around the house in search of the matches so that I could light a lamp. When the first winter came and the days got shorter and shorter, so did my workday. No construction work could be done by kerosene light. My eyes would tire reading under the kerosene lights, and I would go to bed early. I sometimes felt I had regressed to the life of my caveman ancestors.

As I thought more about lighting in the planning stages of this house, I realized that only part of the plan should involve the electric lights of the future alternative energy system. The other part of the plan should consider the natural daylighting of the house. I considered the window placement of the few east side and west side windows. They needed to be located in places that would provide good daytime reading light. The south side of the house had so many windows that all the rooms except for one northwest first-floor room would get ample general lighting. Of course, the sun-space areas of the first floor would be bright on even the cloudiest day. I planned to have very white walls and ceilings in all the non-bedroom rooms.

I have always loved rooms that are bathed in light. I really would like to live in a greenhouse, or even outside. I felt that the sun spaces of the south portion of the house would make me feel as if I were living outside. Now, many years after this house has been built, I have read several articles on the benefits of natural lighting. In our house, or the office above the garage, I always work or read in natural light during the daylight hours. At worst, on a very cloudy day I move to a seat closer to a window.

For environmental reasons, I wanted to build this house of local materials whenever possible. The stones came from this land. Much of the lumber for the final stages of the house, for the garage, and for the old offices for Fowler Solar Electric Inc. came from our land, and was sawed at my neighbors' sawmill. The framing lumber came from Clarence Witter's sawmill or Roger Gunn's saw-mill, both in Worthington. The plywood and the pine boards I bought from the lumberyard were not quite as local, but they were eastern lumber supplied by lumberyards in the bordering towns.

Ethically, I wanted any labor that I needed to be supplied by local workmen. The well was dug by Kirke Henshaw of Chesterfield. The plumbing was done by Darrell Shedd of Worthington, the concrete came from Donovan Brothers of Huntington, the chimney was laid by a friend of Clarence Witter's, and the foundation was poured by Charlie Nugent.

The local workmen, or their families, were people I had known for most of my life. I knew what they did well and what skills they might be lacking. There are always imperfections and misunderstandings, but there were many fewer on this job than is the norm. This was, of course, my house, and my money spent for the wages for each workman. But it was also more than dollars spent and hours worked. Each of these workmen has seen me in the years that have passed. They each built a little piece of my house. I like to think they can remember with pride that they poured my foundation, dug my cellar hole, or sawed my lumber. I hope they can remember that they got paid on time by someone who appreciated their efforts.

CHAPTER 8

The Foundation

MY SISTER SUE eventually divorced Kenny Beach and married Bernie Forgea from Cummington. Bernie and I have many similar interests and values. Bernie surprised me one day just at the time when I was getting the house closed in for the winter. He told me how incredible it had been that everything had gone so well with the project. There had been no great interruptions in construction. I had not hit ledge in the cellar, the well had not been too deep, and I had not had to wait for contractors. This was strange to him. Normally there were unforeseen happenings that created delays or additional expense.

I went home that night and really thought about Bernie's observations. He was definitely correct. There were some ways I could account for some of this good fortune. I had planned carefully, and I had done my research. Nevertheless, good luck had accounted for some of the success. Today, when I think about this good luck, I like to think it was not the kind of luck that one has at the roulette wheel or the poker table. It was more the kind of luck that my father would describe as "holding your mouth right." I never understood exactly what that 1920s phrase meant, but it is akin to shooting a foul shot in basketball and guiding the ball's trajectory after the ball has left your hands.

I feel that my good luck was some sort of return for having tried to do the right things. The Bible has the Golden Rule, "Do unto others as you would have them do unto you." Eastern religions have karma. The Native Americans believe they must live in harmony with nature. I really feel that the majority of the time I worked in harmony with the environment and the people on this house project. For this I was rewarded.

My neighbor, Bob Cook, had often quoted to me, "You have to

The foundation.

have a sense of humor if you build your own house." I always took this to mean that you had to be able to step back once in a while and laugh at yourself, your mistakes, your stupidities, and your shortcomings. I have added a second axiom: "You have to be willing to do things over if you build your own house." You will make many mistakes, and you will always find a better way later. You can always do it over. If you do not waste your energy fretting over the spilled milk, you will probably have enough energy left to redo it.

Zack Donovan had completed digging the cellar hole. When it was two feet short of full depth, Zack started to find very large boulders in the hardpan. His 'dozer was too small for these conditions. He offered a compromise. Zack proposed that he stop there. He could gain one foot if I was willing to have a seven-foot basement instead of an eight-foot basement. He could dig a one-foot trench around the perimeter of this less deep cellar hole for the wall footings to gain the other foot. There were some advantages to this compromise. Later, I would not have to wheelbarrow gravel into the cellar to fill up to the level of the footing before the concrete floor could be poured. Also, the foundation would cost one-eighth less. Zack would finish sooner, so the cellar hole would cost less to dig. In hindsight, I could have hired a bigger 'dozer to dig the full cellar hole. I just did not know enough at the time to realize I had saved some money with the compromise, only to later spend double what I had saved to add fill around a foundation that stood too far out of the ground.

Zack and I needed to map out the final size and shape of the foundation in the cellar hole. Ultimately, four stakes needed to be positioned in the cellar hole such that they formed the exact rectangle of the future foundation. The rectangle needed to measure 36 feet by 24 feet, with four accurate 90° angles at each of the four corners. For the rectangle placing and sizing Zack would normally have used the accepted construction method of trial and error measurements and placement of the stakes until he had honed in on the final placement. He knew that at the end of this trial and error procedure, each side needed to be the correct length, and the distances from opposite corners had to be the same to assure four 90° corners. This was an adequate method.

I couldn't conceive that a better way to stake out the cellar hole, based on geometry, would not be appreciated by Zack. I am sure he gave me a few hints that he did not want to hear my way. For me, my way was so simple. I had calculated the diagonal of the rectangle. I could stake one side and use two tape measures to find the intersection of the other side and the hypotenuse. The whole job would be done in a few minutes.

I did not realize that 60-year-old Zack saw me as a 33-year-old kid who thought I was too smart, This was especially irritating to him at the end of a hard day. My method failed to work that day, as it had worked before, and works today. Eventually, Zack said he did not think much of my clever ways. We finally completed the dimensioning his way, and he happily went home. After he left, I retraced my steps attempting to find my errors. I found where one of my initial stakes had been moved and had thrown off my geometric method. This was probably a mistake made by Zack during a procedure that he was certain was useless.

I learned a valuable lesson while working with Zack. I had to find a way of working with someone who might be resistant to me and my ideas. I do not mean a situation where someone disliked me. On the contrary, I have always had great respect for Zack, and he has always liked me. I believe Zack was confused or threatened by my new way of doing this task. He had done it his way for 40 years. What I learned was to be a little more salesmanlike and humble when I wanted to teach an old dog a new trick.

The forming and pouring of the foundation came next. Prior to digging the cellar hole, I had spoken to Charlie Nugent about pouring my foundation. I told him when I would be ready. Charlie had kept me in his schedule and showed up exactly on time. I was amazed. I had been prepared for delays resulting from workmen who had overscheduled.

Charlie Nugent is about six years older than I am. As a young local kid I tended to know all the older kids, even the kids in the neighboring towns. I had a sister who was four years ahead of me in school. I was privy to her and her friends' conversations of which boys were this and that, and what cars they drove. At local events such as the Cummington Fair, these older kids were on parade at the high stringer or at the cotton candy stand, one arm around a girl, and the other arm supporting the cigarette pack rolled up in a T-shirt sleeve.

The four Nugent brothers and five sisters grew up in the next town, Chesterfield. Several of them now live in Worthington, but I think Charlie was always my favorite. As a teenager he was a wild character, the kind who stood out in the crowd. He drove hot cars that were low in the rear, and he talked loudly with great animation. When I was a freshmen in high school I sat at study hall table with Sherry, a wild junior, who befriended me mostly because I looked up answers for her while she took make-up exams. She was dating this older guy Charlie Nugent and later married him.

Charlie arrived in an old International two-ton, four-wheel-drive truck. The rear axle gears of the truck were blown, so Charlie had pulled the rear drive shaft. He drove the truck in four-wheel drive all the time. Since the rear drive was disconnected, it effectively became a front-wheel-drive truck. This poor old truck lumbered to the job overloaded with old plywood forms. Charlie had taken over Win Donovan's concrete business. He still had the same forms that Win had used for years. The forms were in pretty tough shape at a time when many people in the business were switching to ones that were newer, fancier, and easier-to-use.

Bob and Karin Cook's pond.

Charlie just kept bulling along with the old forms, which had probably been used on the foundations of half the new homes in Worthington over the past ten years. These forms had character, good local character. Charlie was a bull in general. He had a strong, muscular, though rangy, physique as he grew up. After years of construction work he just added more muscle. He also bordered on hyperactive. He could work harder and longer than most anyone I had ever known.

When Charlie went to work, he was a driven man. When forms were needed, he raced the pile off of the truck. Charlie was notorious for driving his workers into the ground. On this job he had only one helper, a quiet and good-natured eighteen-year-old kid. Charlie's other workers had just suddenly found other jobs. Charlie ranted and raved at this kid to keep him moving at a pace still only about half as fast as his own. I was a little worried about this foundation crew. I was not worried about Charlie; he was a good guy who shouted loud orders. The kid seemed to take it all in stride. I was worried about the newest worker and third member on the crew, me, Jeff Fowler. I needed to keep the costs low wherever possible. I had offered to work for Charlie on this job in order to receive a price reduction on the final bill. Charlie knew where I was coming from because he had built his own house. He said yes.

I have always done a lot of construction work. In addition to working at the general store while growing up, I painted houses and those high roofs that no one else wanted to paint. After college, I was too anti-establishment to settle for a career job that utilized my education. I often worked as a carpenter, and as I worked, I watched all the other workers to educate myself in all facets of carpentry. Therefore, I learned carpentry faster than the average person. Much of the carpentry work I had done was in Cambridge and not in Worthington. I knew the local workers like Charlie had to have doubted my skills. Furthermore, I was building a "house of a different color." This meant the workers were being asked to do work they felt was questionable, by someone they doubted.

Theoretically, this passive solar house should have been oriented due south. I was concerned about a margin of error. I decided I wanted the error to be no more than a ten-degree deviation to the west. I carefully compensated for the fourteen-degree magnetic deviation between true north and the magnetic north of the compass. We lined up Charlie's transit so that the compass had the foundation oriented 5° to the west. This would be halfway between 0° and 10°. We double checked with my Boy Scout compass.

THIS FOUNDATION was more complicated than a plain rectangular foundation. Two interior walls would support the 36-foot long suspended slab. There were some places where a portion of the exterior foundation wall would be sixteen inches thick instead of the normal eight inches thick. This would provide the necessary support for the stone wall that would be built above it.

I had a great time working with Charlie. We talked and joked, and since I can also be a driven bull of a worker, I kept right up with him. The ultimate compliment came at the end of the job. Charlie was very generous with the pay he discounted from the job price. He rewarded me for my hard work.

It was common knowledge amongst the other construction workers in the hilltowns that Charlie had one failing. He inevitably ordered the concrete to be delivered too early. A concrete truck mixes the concrete while in transit. When the truck arrives it needs to pour the concrete before the concrete hardens inside and becomes a ten-cubic-yard concrete chunk in a truck-shaped tin can, suitable only for use as a giant anchor.

As one o'clock approached, we were not on schedule. Charlie kept trying to eliminate unnecessary tasks to get ready in time. I was worried. I had helped on enough foundation pourings to guess that we did not have enough deadmen, or braces, to keep the foundation walls from shifting. Charlie was sure we would get by.

Just before the concrete truck lumbered down the road, the cavalry arrived. Win Donovan, the oldest son of Zack Donovan, showed up with a friend during their lunch hour. They were looking for Charlie. They needed him badly for the job they were doing.

Win Donovan was an amazing man. In the past, he had told me how he had been your average contractor around Worthington. He drank beer, he smoked cigarettes, and he ate any food while constantly on the run with his construction business. One day he was diagnosed with an auto-immune kidney disease. Win told me his doctors had recommended a kidney transplant at a time when those operations had had poor long-term success rates. Win's youngest brother, Greg, who was one of my sixties friends, was a macrobiotic. Win, his wife, and his family traveled to Boston to study macrobiotics in hopes of combating his illness.

Win showed up at my place looking a little gray and too skinny, as most macrobiotics do, but he was alive to save the day. (Win lived for a good ten years longer than his doctors had predicted.) Win and his friend quickly sized up the situation. They skipped lunch. Now our three-man crew was five. The concrete truck arrived. When we started pouring, the walls started moving. This was no problem with our five-man crew; we quickly dropped in more deadmen where needed and restraightened the walls. The day was saved. The foundation was good. Win was happy because he got Charlie off my job and over to his job.

During the first weeks of construction, I found it necessary to make a few improvements in my standard of living. The first improvement was the purchase of a VW poptop camper. I bought this from a friend who could not easily sell it. It was old, and tired, and rusted. However, the fold-down bed, the table, and the interior were in new condition. I paid $350 for my luxury. (When fall came I sold it to another friend for $350.) I now had a nice dry place to share with my little dog Rolo.

The camper had no refrigerator so my food supply was still catch as catch can. Before the foundation was completed, I buried a plastic bucket and lid in a hole in the ground for food storage. The ground temperature was 60°F. Foods like yogurt or tofu would keep for a few days. I usually made a trip to the general store each day. Often I would finish work after the six-o'clock closing of the store. On these days I would ravenously drive twenty miles to the city to stuff myself at a restaurant and buy materials and tools from the stores that were open. My standard lunch was a cold refried bean sandwich with taco sauce and mayonnaise. I was quite happy with my lunch, but sometimes I suffered some embarrassment when a worker saw me eat this same lunch day after day.

The weather was getting warmer. I had become better friends with the Cooks. They offered me the use of their large pond in the woods for bathing. I was rapidly closing in on suburbia. When the cellar was finished I kept food on the cold 50° F. floor. There were disappointments, like the day when my friend Jonathan's dog found his way into my cellar food storage while Jonathan and I were talking. After Jonathan had left, I found the damage and faced the prospect of a night without dinner.

My next improvement was a telephone. The nearest pay phone was five miles away. I had been wasting increasing amounts of time getting to a pay phone. It also became clear that I desperately needed to be available for callbacks. The phone company would only install a phone indoors. As soon as the cap was on the foundation, I had the phone installed in the cellar. There were some comical scenes for my wildlife friends over the next few months. When the phone rang after I was tucked in bed in the camper, I would dash naked and barefoot 100 feet to the cellar to answer it. While I spoke, I fed the mosquitoes.

To order the telephone I called the appropriate number listed in the beginning section of the Cooks' phone book. I spoke to a nice woman and ordered the phone. At the end of the conversation she politely announced in her nicest operator-type whine, "Your new service will be installed in three working days." I tried to explain that this phone service would need more attention than a regular installation. She simply could not conceive of the conditions of my location. I finally gave up and hoped for the best. After all, she had been polite and helpful. She probably had seen places like Bashan Hill only in Walt Disney nature movies.

The salesperson at the phone company was sure the installation man would arrive and run a wire for 50 feet from the pole to the house. In this case the house was more than 200 feet from the road. The pole crew would have to set a pole 150 feet into my property first. To make things worse,

Bashan Hill Road from the Cooks' house to my land.

there was no available telephone line on Bashan Hill Road at the end of my driveway. The only telephone wire was a dual conductor cable suitable to service one telephone. This line went to Phil and Ann Taylor's camp down the road. The nearest multi-line cable was 1.3 miles away, back up on Route 143. I actually felt very guilty about all of this. For a thirty dollar installation charge, I was making a lot of work for the phone company.

A telephone service man showed up within the three business days, right on schedule. He looked like he would have felt more comfortable if he had packed a compass and an army survival outfit. He smiled and quickly left. The next man to appear was the district engineer. He told me it would take a little longer to complete my service installation. Two weeks later a line crew ran a single service wire from Route 143 to my door. Half of the standoffs used to hold the wire were on telephone poles. The other half were screwed into "adolescent" telephone poles — live trees with branches and green leaves on the tops and roots in the ground.

WHILE I AM ON THE TOPIC of the freebies and surprises associated with getting 1.3 miles of telephone wire for thirty dollars, I will tell you about my road improvement, though some of the story will be a month or two further into the summer. Bashan Hill Road was a rough road, but it was a town road, meaning that it was graded, raked, and crowned in the summer and plowed in the winter, all by the local town road crew. There were no houses after the Cooks' house. Therefore, the road was minimally maintained after that point. The last one-third-mile stretch to my driveway was graded, but the gravel content was very low. It became a sinkhole during mud season. When I purchased the land, I planned to park at the side of the road near the Cooks' and hike in the final distance during mud season.

Jim Pease was the highway superintendent, and Ernie Nugent and Gerry Bartlett were the crew. Worthington has had the best road crew in the hill-towns for years. When I first took possession of the land, I had stopped down at the town shed sheepishly looking for Jim Pease. I felt guilty asking for favors because the out-of-staters were always putting pressure on the town's resources by demanding more services from the town police and the highway department. Demands like, "Can you come sand my road? It's Sunday night and I have to get back to the city, and my car (not equipped with snow tires) can't get up the hill" were not uncommon.

I was sure the town did not have the money to instantly upgrade my road. I did want to talk to Jim Pease so that my road could be put on the long list of needed improvements. I hoped that my road would come up for gradual improvement over the next three or four years.

Jim Pease gave me a big hello the day I walked into the town garage. I grew up knowing the four or five Pease families, Jim's family in particular. Ernie Nugent was Charlie Nugent's brother. I went through grammar school and high school with Gerry Bartlett. I was not walking in and looking for favors from friends. I was patient. I was aware of the budgetary constraints for road improvement in our small town.

Before I could apologize my way into asking for some road improvement in the future, Jim Pease told me he had been meaning to stop by Bashan Hill Road to see me. He had heard I was building a home there, and he was sure I would need some improvements. He said he had some money allotted for gravel on my road. He also might have some excess gravel from some other road jobs to throw in. Jim doubted that he would have enough gravel to repair the road all the way to my driveway, but he would come close this summer and reach my driveway the following summer. I was astonished. One of the real liabilities of my new land was instantly solved. I left feeling grateful that I was a native son.

The road crew made it to Bashan Hill Road in July. They added gravel to some trouble spots on the hill. Somehow they found enough to add six inches of gravel all the way to my driveway. The road has been in good condition ever since. On different days during the summer each member of the crew found a few minutes during a break or a lunch hour to stop in and see how the house was progressing and who was working there that day.

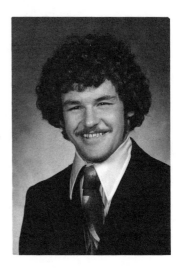

My nephew Scott Beach.

I NEEDED A GENERATOR in order to build the house. I had decided to purchase a top quality generator and use it for several years until my alternative energy system would be in place. Bob Cook suggested that I buy a Miller welder-generator. The quality of the generator was good, and I would basically get a welder for free. He sent me to Merriam-Graves Corporation in Pittsfield. Merriam-Graves had a new salesman who had recently been promoted from driver: the same Butch Parish who had given me my first ride down Bashan Hill Road in his jalopy. Butch gave me a good deal on an 1800 RPM electric-start welder. This model was also a powerful commercial generator. Butch had married my sister's best friend from high school, Betty Brooks. He was one of my nearest neighbors back on the main road, next door to the house where he was raised. He even delivered the six-hundred-pound generator and helped my nephew Scott Beach, Charlie Nugent, and me unload it.

Scott was eighteen years old in the summer of 1981. He was just graduated from high school. He had been accepted to a junior college in Colorado for the fall semester, and he was looking for a good summer job. Scott and I have always been very close. He has always been more like my younger brother than a nephew. We had worked together on his mother's house when he was a precocious ten-year-old child worker. Most of all he was enthused about the house.

Scott and I and the house project were a good match. Scott wanted this summer job to earn money for college. The previous summer he had worked clearing roadsides for the town of Cummington. It was backbreaking and brainless work. The house project would be hard work, too, but it would be a new experience every day. I paid Scott more than he had expected to be paid because he was worth the extra pay. I could not afford a real carpenter. I also wanted a helper who would follow the new methods of construction I would employ for this energy efficient house.

One difficulty in the early summer was Scott's health. He had some non-descript stomach ailment or irritation. Many times he came to work in the morning, and I had to stop work and drive him home at noon. My problem was that I had too many deadlines. For the first few weeks Scott only worked 30 hours instead of 40 hours. My solution was to add ten hours to my schedule. This would have been acceptable if I had been working 40 hour weeks, but I was working 70 hours a week, then upping it to 80 hours per week.

Scott and I had a few bad days together but not many. I was inclined to push too hard. During some weeks, I was completely caught up in this project for all my waking hours. Some days it was only a job for Scott. Most of the time Scott worked as hard as he possibly could. He still feels like this house is his house. He loves the setting, and he loves the design. I can still hear him say that the cloud formations here are the most amazing he has ever seen. I was so busy that summer that in many ways Scott was my only friend.

CHAPTER 9

Building the House

THE FOUNDATION was finished, and Charlie's front-wheel-drive, almost four-wheel-drive, old truck had waddled off, overloaded with old plywood forms. Scott had survived his hectic party week and high school graduation. We installed the drainage around the foundation and waterproofed the walls. Clarence Witter delivered the framing lumber, and Bisbee Bros. delivered the plywood. We were ready to start building the house.

We chose to pour the floor in the cellar before we began framing the house. Most contractors in our area pour the cellar floor, after the foundation is capped or after the house is closed in. The advantage of pouring the floor later is that rain cannot fall on the floor while it is setting up. Raindrops on a setting concrete floor spoil the smooth finished surface.

We planned to pour the floor on a clear day, but we were also prepared to cover the whole foundation with polyethylene in case of an unexpected shower. There were real advantages to pouring the floor first. The concrete truck would be able to circle the foundation and send the concrete to all parts of the basement. This would mean less backbreaking shoveling and no wheelbarrowing. Because this house would have a first-floor suspended slab, I felt it was advantageous to pour the cellar floor first. The basement concrete floor would be a level supporting surface for the 2x4 posts that would support the heavy weight of the suspended slab pour.

Every house has its mistakes. As I have said before, you have to take solace in the fact that you can always go back and repair them. The cellar floor was the one part of the house project that went poorly. It took me years to repair the error. I was prepared to pour the floor the correct way. The day of the pour I asked the wrong question at the right time, and consequently the floor was done the wrong way. I feel that I got poor information from my friend who was helping me pour the floor. Free or traded labor is rarely as responsible or as professional as paid labor.

I told Charlie that I would not need him to pour the cellar floor because my friend Pierre would be the boss on the pour. This would save me several hundred dollars. I had helped to pour many cellar floors. All I needed were a few strong backs and someone with the expertise to manage the pour. Pierre wanted to trade his time overseeing the pour for my time on a roofing job.

Pierre arrived early in the morning to prepare for the pouring of the floor. Scott, Pierre, and I were planning to grunt our way through the job with no other help. I had been determined not to use plastic on this project unless it was absolutely necessary. It was an environmental choice for me. Everybody who pours floors knows that you need polyethylene under a cellar floor to prevent moisture coming up through. Zack had mentioned it. Charlie would have used it. I had polyethylene there ready to be installed. I made two mistakes. The first mistake was that I wanted to be a purist. The second was that I asked Pierre the big question: "Do I really need polyethylene under this floor?"

When I had worked on a few carpentry jobs before and after this with Pierre, someone would inevitably ask if a cut, or a fit, was good enough. His stock country witticism was, "You'll never see it from my house." Pierre told me that the polyethylene was not really necessary. He didn't use it in his basement, and he didn't have a problem with moisture coming up through the floor. That was the answer I wanted to hear.

Pierre, Scott and I poured the floor. Pierre left while the cellar floor set up. Scott and I found other work to do. Late in the afternoon Pierre returned, and we all troweled the hardened floor. We had all worked exceptionally hard. As a reward, we went for a swim down at the Cooks' pond. While swimming and drinking a beer, Pierre said, "I got to go. I got a softball game. I'm sure glad we didn't use the polyethylene. Those floors take forever to set up. We would have had to wait longer before we could have troweled the floor, and I would have missed my ball game."

Pierre had a reputation for working hard when the owner was at the job, and for getting stoned for a long break if the boss was gone. He could be almost your best friend and then subtly slip into his "me first" perspective. I was worried about the floor the moment I heard about the ball game. I still give Pierre the benefit of the doubt. He may not have been as knowledgeable about cellar floor design as he thought. He may have believed that the use of polyethylene was not always necessary. I do know the first thing in his mind was the ball game.

I knew the polyethylene should have been there. Zack had not dug ditches for me to put drainage pipes and crushed rock underneath the floor, because the base was fair gravel, and because he had double-checked with me that I

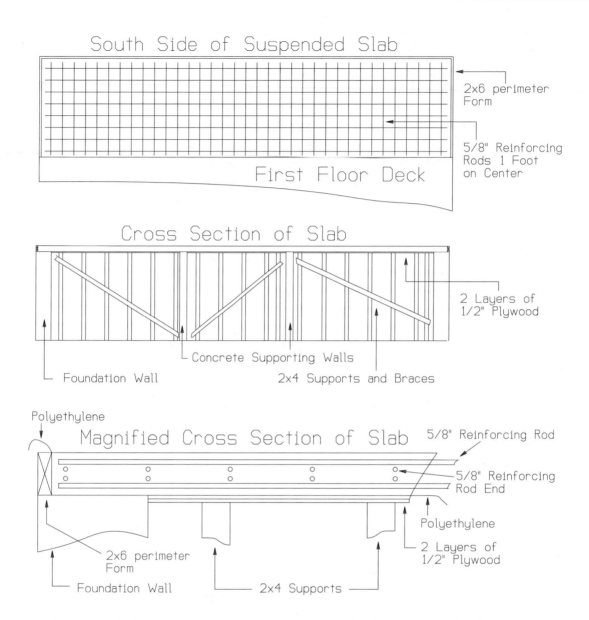

South Side of Suspended Slab

2x6 perimeter Form

5/8" Reinforcing Rods 1 Foot on Center

First Floor Deck

Cross Section of Slab

2 Layers of 1/2" Plywood

Concrete Supporting Walls

Foundation Wall

2x4 Supports and Braces

Polyethylene

Magnified Cross Section of Slab

5/8" Reinforcing Rod

5/8" Reinforcing Rod End

Polyethylene

2 Layers of 1/2" Plywood

2x6 perimeter Form

Foundation Wall

2x4 Supports

Details of the suspended concrete slab.

was, of course, planning to install polyethylene under the floor. Another problem with a pour that is not on polyethylene which was not known to most workmen at the time is that the liquid of the concrete leaches out into the sand or gravel below. This weakens the concrete. What Pierre did know was that the floor would set up faster without the polyethylene.

For the next twelve years I lived with a cellar that wicked moisture up from the ground below. Moisture was trapped under anything that was placed on the floor. In corners, and where the walls meet the floor, the reduced air circulation left the wicking moisture on the surface. My cellar was always damp. I tried coatings and paints. The hydrostatic pressure from beneath always forced the paint or coating off the floor. Finally, twelve years later, I put down a sheet of polyethylene on the existing floor and poured an additional thin slab. I did this in sections using an electric cement mixer. My foolish question in 1981 had caused me 12 years of inconvenience and three back-breaking weeks of work to repair the problem. I still take solace in the fact that ultimately all I had to do was just do it over.

OUR FIRST HOUSE BUILDING JOB was to place the carrying beam and its supporting lolly columns, and to cap the foundation. In this case, we only needed to cap about two-thirds of the top of the foundation. The front one-third of the first floor would be a suspended concrete slab. Capping a foundation goes quickly. It is easy to mark and lay out the joists. All of the plywood nailing takes downward strokes of the hammer. Scott and I loved it. We had the supporting beam in place and the foundation capped in two days. We felt as if the whole house might be finished at the end of the week, and we could have the whole summer off.

Our next job would be to make forms for the suspended concrete slab. The whole slab would be nine feet wide and 36 feet long. Charlie and I had formed two extra foundation walls interior to the basement to support the 36 foot long slab at 12 feet and again at 24 feet. Three table-like forms supported by the cellar floor seven feet below would form the horizontal base for the slab. The south, east, and west edges of the slab would be formed by 2x6s. The north edge of the slab would be the box sill of the first floor deck. All of these forms together would create a six-inch deep, nine-foot by 36 foot box to be filled with concrete.

Braced 2x4s supporting the forms for the suspended concrete slab.

I WAS WORRIED about the forms that would support the slab, which would weigh more than 70 pounds per square foot. If this form sagged over the first week while the concrete was hardening, the concrete would develop cracks. I was always secretly afraid that I might botch the job, and that someday the whole slab would fall in, crushing me in the pile of rubble. This was, of course, absurd. I had Mike Lucey's professional engineering plan for the reinforcement of the slab.

Scott and I cut 2x4s to length such that one end stood on the cellar floor and the other end would support the plywood floor under the slab. We placed an upright 2x4 every two feet in all directions. It looked like a 2x4 jungle in the basement. We diagonally braced all of these 2x4s together to prevent them from bending under the weight. On top of the 2x4s we placed a floor of half-inch plywood. On top of this, we placed another layer of half-inch plywood such that the joints of the two layers never matched up in the same place to create a weakness. On top of the plywood and folded up on the insides of the perimeter form, we placed a continuous sheet of polyethylene. This polyethylene would keep the plywood clean for reuse, and it would keep the liquid component of the concrete from seeping out. It is the water-and-cement liquid that becomes the glue which bonds the sand and crushed rock together to form concrete.

We now had a six-inch-high box that was 36 feet long by nine feet wide. We next needed to add a network of steel reinforcing rods that would provide the tensile strength to the slab. Mike's design called for a bottom grid of rods that would sit one inch above the bottom of the form and a second grid of rods that would sit one inch below the top of the finished slab. Each grid would consist of 5/8 inch rods whose network made one-foot squares.

Bob Cook came to install the reinforcing rods. We had agreed that I would later repay him with my labor. He brought the tools and the know-how of an experienced ironworker. We used up the excess reinforcing rods and thus overbuilt Mike's design. This slab would support a truck.

That night I ordered the high strength, 4000 PSI concrete for the slab. We planned to have plenty of labor for the pour. Scott and I would be assisted by Gerry and his friend Roger. Gerry arrived on time, but without Roger. Bob Cook showed up unexpectedly, so we luckily had plenty of labor. All went well. The driver of the concrete truck was another of Zack Donovan's sons, Corky Donovan. He worked for his uncles at Donovan Bros., a local sand, gravel, and concrete company. Corky was a friend from grammar school. He was very patient with us. He expertly placed the heavy slump concrete, saving us much shoveling.

Concrete does not set in a day. The process is actually a ten-page chemical equation that proceeds over years. The majority of the setting of concrete occurs in the first week. The surface of concrete must not be allowed to lose water by evaporation. We placed a sheet of polyethylene over the top of the

The concrete slab curing under a layer of polyethylene covered with sheets of plywood.

slab. For the first week, we used a hose to keep the surface of the slab under the polyethylene moist. I was so over cautious that we covered the slab and polyethylene at the end of the first week with sheets of plywood so that we could work on the surface while we let the slab cure for two additional weeks. At the end of three weeks we dismantled the forms. We saved all of the plywood and 2x4s to use in building the house.

At this stage of the construction I should have insulated the outside of the foundation and had the foundation backfilled. When Zack was digging the cellar hole, I had told him that my distant relative Alan was out of work and that I had arranged for him to help me. When Alan was called back to work, I was left with the crow-eating experience of calling Zack back in the peak of his busy season. Zack was pleasant about the change of plans. He would be over as soon as possible, which would be a few weeks. He wanted to come and backfill at the same time he did the septic system. This would save me money because he could work grading while waiting for materials and deliveries of septic system parts.

Zack had been so nice to me that I did not have the heart to push him to come sooner. Our problem was that we were starting to build the house while the house sat in a hole that surrounded the foundation. We could not stand on ground level and work on the first floor walls. When we got to the second floor and Zack still had not come, we had the inconvenience of effectively working on the third floor. Every piece of material had to be carried up an additional height of eight feet. Scott and I just ignored the inconveniences and kept building.

One of the biggest conveniences was the generator. We constantly used circular saws, a radial arm saw, and a table saw. One of the biggest inconveniences was the same generator. This was a great generator as generators go.

I was gradually learning that I basically hate generators. Here we were in the peaceful wilds listening to a loud generator. We could not stand the noise all day; we left the generator running only when necessary. As the house grew, it became a workout running up and down ladders to start the generator.

The house kept growing steadily each day. It looked absolutely gigantic, perched up on the exposed foundation like a big bird. Scott and I had grown to enjoy the outside emphasis of an unfinished house without really knowing it. We would break and lunch on the suspended slab in front of where the windows would be on the south side. We looked at Bashan Hill to the east and the beautiful cloud formations that are more evident at higher elevations. We were really enjoying our progress, and we appreciated the remote location.

Normally the only vehicles on the road were delivery trucks. Occasionally a car would drive to the house site. Someone from town would come for the twenty-five cent tour. When we would hear a vehicle, we would stop all work to determine if it was a delivery truck. Often we ran to the road to direct a driver who had not yet delivered to us. We developed a game for the times when it was just a car or truck passing through. Scott and I would both yell in unison at the top of our lungs, "Who's driving down *my* road."

At one point during the summer a twelve-year-old kid was staying with his dad in a camp a mile away on Parish Road. He spent all day, every day, racing up and down Parish Road and Bashan Hill Road on an old and noisy dirt bike. Finally one day, he came down the hill as fast as he could as my sister was driving up the hill after having dropped Scott off for work. The motorcycle kid scared my sister badly, but she managed to stop as he laid the bike down in the dirt.

The next day I heard him coming. I decided to be the enforcer. I ran out the driveway, stood in the middle of the road, and waved both hands. The little bugger nailed it and went right around me. I guess I was not too intimidating. Two hours later he was back. This time Scott and I ran out together. I gently motioned for him to stop while I leaned on the four-foot 2x4 that I had brought along. The kid stopped. I asked him for his motorcycle license. Of course, twelve-year-olds do not have licenses. I told him this was a town road: no license, no motorcycle. Peace returned to the wilderness.

When it came time to put up the rafters, Scott and I procured some extra help. My friend Jeffrey Hartman, an oil painter from Boston, drove out for a few days to help. He was not even close to being a carpenter, but he was one of my best friends and one of my only friends to volunteer on the house project. Scott's friend Willie Brown was hired as the fourth person.

The rafter design for this saltbox was fairly complicated. The north side of the roof was supported by 30 foot long rafters. Each 30 foot rafter consisted of a 16 foot 2x6 and an 18 foot 2x6 that were overlapped four feet and nailed together. The overlap of the two parts of the rafters sat above a long wall that

Building the house before Zack arrived to backfill the foundation.

would be the front walls of the upstairs closets. I was the only carpenter on the job, though Scott was learning fast. I attempted to plan the rafter raising such that I could be one carpenter with three helpers, two of which were inexperienced. We got the ridge pole up and the rafters cut and in place in two days. We braced our construction at the end of the day. The job was well done. Friday night arrived, and all my helpers went home.

Early Saturday morning I was making the rounds and double-checking things before I started my weekend solo work. I put the level on the second-story south wall and discovered it was out of plumb. The whole wall had pushed out an inch at the top. I knew that it had been plumb before. As I looked around, my giant mistake became very evident to me. I had not built the wall that should have supported the back rafters. I had been one carpenter with three helpers. I had braced the rafters and planned to build the supporting wall myself on the weekend. Simply put, the braces did not do the job the wall would later do. The rafters had moved; the front wall had kicked out at the top; and the ridge-pole had sagged. I wanted to cry.

I came back to my axiom that I could just do it over. The only problem was that the only way I could see to solve the problem was to actually do it all over. After hours of thought, I guessed that, if I could pull the top of the south wall back into plumb, then all the other parts would also be pushed back into their previous positions. I attached my comealong to the south wall and tried to pull it back. The one ton comealong was too small.

I remembered that Bob Cook had the biggest comealong I had ever seen. I borrowed Bob's comealong and pulled the wall back, section by section. I worked all day. I was exhausted, but I was also determined to have the whole problem fixed before anyone showed up on Monday. It would be easier to

Raising the rafters.

admit my large mistake in the same breath as the explanation of how I had fixed it. On Sunday I built the supporting wall. I also decided to plywood sheath the east and west second-story wall. This would brace these walls.

My ladder was too short, and I had no staging. To sheath these walls I sent a 4x8 foot sheet of plywood out between the studs, and held it in place outside of the wall, while I nailed it leaning out of the wall from the upstairs floor. The 50 pound sheet of plywood got increasingly heavy the farther I pushed it out, before I could turn it. I mostly killed myself. I dropped every second sheet of plywood I tried to install by this method. I let go just before I was forced to take the long trip down with the plywood. It was a miserable weekend of work.

I was proud of one of my rules on the job. As the very end of the day approached, I made us get more careful. I remembered as a boy how the ski-patrol toboggan carried most of the injured down the ski slope at the end of the day when tired minds and bodies made more mistakes. Scott always resisted this rule. His macho self-image wanted to "go-go" at the end of the day. During the last half hour I would remind him to be careful on the ladder and with the power tools. Scott was like my little brother. I could not conceive of him getting maimed building my house. One late afternoon when he was reckless on the roof, I relegated him to the status of gofer for the last half hour. He was red in the face with anger, but he was safe.

At the same time I was the carpenter building the house with Scott, I was also the general contractor on the job. I ordered the materials and scheduled the work of the outside contractors. This work was normally done in the evenings and in the early morning before Scott arrived. Each day of work had to be planned in advance. Without the planning, the job would stop for lack of some tool or material.

While Scott and I were capping the foundation and building the forms for the suspended slab, Kirke Henshaw of Chesterfield was drilling our well. Kirke now runs a good-sized well-drilling business. At this time Kirke had only one well-drilling rig and one part-time employee, his neighbor. The drilling rig was an old one. This old style drill hammered as it drilled. Scott and I hollered above the deafening noise of the hammering for two days.

It is easy to see why Kirke's business grew. He was good at the drilling, and he was honest. Much of the expense of the job for a well driller is the set-up time of the drill before the drill even begins to turn. The actual amount the well driller charges for the well is the product of so many feet down at so many dollars per foot. It is easy to see that a well that yields water at too short of a depth could cause a job to lose money, while a very deep well could bring in extra profit. It is profitable for a well driller to drill a well deeper than is necessary. Kirke did his drilling by the book.

Even today, many people prefer to have a well drilled by the old style hammer drill. Our aquifer in the hilltowns, in all but a few river valleys, is held in the fracture porosity of the bedrock. Our local bedrock is never completely solid. The network of interconnecting cracks is the container for the underground water. A rotary drill bores until it intersects a sufficient system of cracks to yield enough water. The hammering drill sets up additional cracks in the wall of the hole it is drilling; this increases the chances of hitting an established system of cracks.

In 1981, Kirke offered to drill my well for $7.50 per foot. Our contract was non-existent by most people's standards. I met him for the first time when he came to the well site. Kirke was well recommended to me. He told me what he charged per foot. I trusted this price would not change, and he trusted I would pay him promptly when the well was completed. This was the way all of the jobs were done on my project. The only written and signed document that was required during the whole project was the deed to the land.

The drilling went uncommonly well. Drilling a well is always a gamble. I was fretting over the prospect of an expensive well. Unlucky people have had 400 foot wells dug. Very rarely people have dug deep wells and then had to abandon the dry hole to dig another. I was hoping for a well less than 200 feet deep. This would fit into my budget. Kirke hit water at 40 feet. It was a marginal yield of water. He advised me to let him dig deeper, lest this supply be too dependent on seasonal water levels. Kirke drilled farther until he hit additional water at 80 feet. The maximum yield they bother to record for a home well is 15 gallons per minute. My well greatly exceeded that rate.

Kirke was helpful with my choice of a deep well pump. The normal installation would have used a 1/3 or 1/2 horse power 240V pump. I wanted a 1/3 horsepower 120V pump. This smaller 120V pump could be powered by my generator now and hopefully by an alternative energy system in the future. Kirke had installed a few of these pumps at remote sites in the hilltowns that pumped water with a small generator. We had one problem on the well project. Kirke's assistant had mistakenly looked up the wire sizing for a $\frac{1}{3}$ horsepower, 240V pump instead of for a $\frac{1}{3}$ horsepower, 120V pump. When I discovered the problem, Kirke immediately responded. He and I pulled the pump and changed the wire. I happily paid the difference for the heavier gauge wire.

I changed the normal installation of this pump system. I substituted an old style galvanized pressure tank for the currently used bladder-type tanks. In any pressurized tank, air from above pushes down on the water in the tank to maintain the pressure in the water system. The bladder-type tanks have a membrane between the water in the tank and the air in the tank. The older tanks actually have compressed air directly above the water level in the cylindrical tank. The air above the water in the old-style tank dissolves into the water and is carried away. Yearly maintenance is needed to replace the lost air.

I was willing to do the maintenance associated with the galvanized tank because the old-type galvanized tanks came in a much larger 120 gallon size. I planned to power this pump by the generator for the first year. I would be able to pump up the large supply of water and not run the generator until the following day. I would not have to listen to the generator during a peaceful bath in my sunny solar bathroom.

As soon as all had gone home on the day the pump installation was complete, I attached my garden hose and nozzle for my first Bashan Hill shower. The deep well yielded cold water that was the same temperature as the cold brook. Before I ate dinner that night, I perched an old hot-water-heater tank on the deck of the first floor. I attached a short hose to the bottom drain faucet. I filled the tank with water through a top pipe hole. The next day this black tank basked in the sun. That night I stood at the base of the foundation in the cellar hole and let the warm water clean me.

CHAPTER 10

I Hate Generators

I HATE GENERATORS. I have probably always hated generators. Cars I can deal with, but small engines, though simpler,
have always been more difficult for me. My first truly bad
experience with a generator was on a construction site in the
late fall of 1972, long before I started this house in 1981. I
was living in Worthington, working for Bob Bartlett, building a log
home on a road where there was no power. I was hired because I was
an old friend of the Bartlett family and because I had a strong back.

The yellow pine logs for this house were newly cut, not kiln-dried,
and therefore were extremely heavy. The longer ones weighed about
400 pounds. These heaviest logs had to be lifted and placed in position by Bob, Roger, Paul, and me. This meant 133 pounds per man.
You may think that my math is ailing, but in fact it was Paul who was
ailing. Paul was an old local boy. He was probably not much older
than the 46 years that I am today, though he seemed more ancient
then, when I was still 24 years old.

Paul was one of many children in a poor family. I doubt that he
had more than a grammar school education, but he was a lot more
savvy than this young college graduate. When Roger, Bob, and I
struggled to place one of these heavy logs six feet up on a wall, Paul
grunted along with us, but hardly lifted. It took a long time for
Roger and me to understand why the large logs that we lifted with
Paul were never much lighter than the ones we lifted without Paul's
help. Paul was smart enough to rest his tired body and let the young
bucks strain their backs. Paul's work was an early lesson in the art of
conservation of energy.

Paul was also quite a wit. About once a day he would come out
with some real clinker, laced with a few creative cuss words, that
would keel us over and send us to the woods before we peed our
pants. Paul spoke in a slow, painful backwoods accent that would be

perfect for shaping Garrison Keillor or Robert Frost quotes. The main reason Paul has found his way from my memory into this book about electricity is because of one of his creative slurs.

I had never thought a lot about utilities and the cost of heating with one type of fuel versus another. I had gone from parents to dormitories to apartments. Our coffee-break conversation around the campfire in the snow outside of the log shell wandered from beer, to deer season, and finally back to the home we were building. Whenever we talked about the log home, it was always in negatives: how badly the logs fit together, how overly expensive it was, and how stupid the owner from New York was for making all these bad choices.

This day Roger wanted to know what type of furnace the owners would use to heat the log house. Bob, the boss, was excited about the new kind of heat, electric baseboard. Paul ended the topic with one of the most succinct and accurate energy analyses I have ever heard. Paul said, "Jesus to Jesus, my goddamn brother-in-law got that goddamn electric baseboard heat. Whenever he turns up the goddamn thermostat, it would like to twist the goddamn meter right off the goddamn wall."

I hate generators. They never seem to be adjusted properly. If your car runs poorly, it does not go up the hills easily, and you are likely to take it in for a tune-up. The average generator gets no attention until it ceases to run. If the generator is on a construction job where someone tinkers with it just enough to get it running, it never gets the tune-up needed to make it run properly. The generator receives just enough attention to get it running poorly. The fuel does not get burned most efficiently and the exhaust smells. Whenever the exhaust smells, you can be sure the pollution level increases, affecting all who are working near the generator.

I also hate generators because they are noisy. At best, they sound like a lawn mower. As the generators age, the baffles inside the small, cheap mufflers start to rot away, and they become even louder.

In late November, as early winter closed in, we finished the shell of the building and installed the windows and doors. One day, our little generator, which began the project running on the first-floor deck of the house, was no longer on the outside, but inside, surrounded by the log house we had built up around it. I hated the smell of the exhaust and rankled Bob's good nature by insisting on working outside with Paul instead of working upstairs with Bob where I was needed. Roger stayed with Bob. The next morning at work Bob and Roger were sick. They had the worst hangover-type headaches one could imagine. They had suffered minor carbon-monoxide poisoning.

The Miller generator-welder that I purchased from Butch Parish at Merriam-Graves was the Cadillac of generators in its category. Although nominally rated as a 5000 watt generator, this one was much more powerful than the common 5000 watt portable generators for homeowner use. My

The Miller generator-welder.

model had an electric start option. It was powered by an Onan two-cylinder engine. The whole unit weighed 600 pounds. It was a beautiful royal blue. At first, I loved this generator, the first real toy that I purchased for the job.

The honeymoon was over with that generator after the first day that Scott and I had to listen to it run all day while we cut studs to length on the radial arm saw. This was a relatively quiet commercial generator, but its noise was still a great invasion of our quiet backwoods site. Scott and I decided to deal with the extra trips to and from the generator to turn it on and off, rather than to listen to it all day long.

Scott was the first to incur the wrath of the generator. Generators are a little like Doberman pinschers; one has to always watch out for the day when one's defenses are down and the unexpected occurs. Scott and I had to slide the 600 pound generator into a different position. We had forgotten that the generator had been running for an hour before we had shut it off, only five minutes before. We both hunkered down and gave a mighty heave. Scott got the end of the generator with the mufflers. One hot muffler branded his left biceps.

When we closed in the shell of the house and shingled the roof, we carefully slid the heavy generator down a steep ramp into the basement. Later, the plan called for the generator to be installed in the basement for easy access. The exhaust would be sent out a pipe through the cellar wall. For the first year or two I would keep the generator in the basement, where I could easily start it to pump up the water tank or power the tools that I would need while I worked on the inside of the house.

The generator was in the basement and working long before the windows and doors were installed. There was no need to vent the exhaust system out the cellar wall because the cellar doors were always open, and the air passed freely through the window and door holes in the house.

Late one afternoon while I was working upstairs, the power quit while I was operating the circular saw. I could still hear the generator running in the basement. I went to find the trouble. All of the breakers were in place in the generator. The receptacle out of the generator powered a drop light. This meant that the problem was in the extension cord. I pulled the plug and noticed that it was damaged. It took me about 15 minutes to replace the plug and another few minutes to test it.

Perhaps you noticed that I never mentioned that I had turned the generator off while I fixed the plug. I had not. I had forgotten. I neither noticed nor cared that the cellar door was closed. I had worked outside for so many days with the generator that I had forgotten the dangers of the exhaust. I finished work, ate dinner, and went to bed.

The morning after fixing the extension cord, I woke up deathly ill. I got out of the camper and upgraded my condition to walking dead. My head hurt, my stomach was nauseous, and all of my senses seemed on half power. Scott had left work earlier in August for a backpacking trip in Colorado, so

I was alone except for Rolo. I went back to the camper and moaned in my bed. I just lay there. I knew no one would be coming that day.

A few hours later a truck backed down the driveway. The truck was loaded with chimney blocks and chimney tiles from Cummington Supply. Peter Kipp and a helper jumped out of the truck. I forced myself to get up and out of the camper.

Peter and his wife, Fran, owned a "ma and pa" building supply store in the next town. Early in my house project I had stopped in for some small items while I was passing through. When Peter and Fran heard I was building a house in Worthington, they quickly, and very politely, asked if they could bid on any of the materials I would need. They felt they could beat the prices of the competition. Peter and Fran were exactly my age. They were out of the same sixties generation. We hit it off well, and they became my friends as well as my suppliers.

Peter knew me pretty well by this point of the summer. He took one look at me and saw that something was very wrong. He was more convinced when I tried to help him unload, as I always had, and I just could not do it. I told him I was sick. I apologized for sticking them with unloading the heavy chimney supplies. When the unloading was complete Peter offered, then insisted, on driving me to the medical clinic up the road. I now realized I was not in good shape.

I called the Worthington Health Center. My sister, a receptionist at the clinic, answered, and I asked for an appointment. I sounded so ill that Sue offered to leave work and pick me up. I managed to get to the clinic in my own car.

My doctor and friend, David Katz, was not working that day. I was sent in to see the new doctor. I had never seen him before. He was quite friendly and kind, but also quite strange. The nurse had taken my temperature. It was normal. This new doctor told me this ruled out the flu. He politely asked me about substance abuse. I told him I was a card carrying vegetarian and health nut. From this point on the diagnosis got stranger. He honed in on the symptom of a bad headache. He was sure I had a migraine headache, and that it was also making me nauseous.

I was pretty slow in my thought processes at this point. However, it did dawn on me that I had worked in the basement with the generator fumes. I threw this information out to the doctor, but he would not hear it. It had to be a migraine. He asked me if I had ever had one before. I did my best to explain to him that I never got headaches at all. He decided this was my first migraine. He then got psychological and asked long questions about mental stress and strain in my personal life. I was so foggy I kept trying to answer. At some point my head cleared enough for me to realize this character was not going to help carbon-monoxide poisoning.

As politely as possible, I thanked the new doctor (no sense stepping on

Barking Dog Camp.

toes in a small town) and took my mythical first migraine headache out to the receptionist desk and my sister. I paid the fee and went home to rest my carbon-monoxide migraine.

I later installed the Miller generator in the cellar. I removed the mufflers from the motor and attached an exhaust pipe system that passed through a hole I had formed in the foundation wall when the foundation had been poured. The system sent all of the exhaust fumes outside, and most of the exhaust noise. I was extremely disappointed with the installation, however. I found the engine of the generator to be mechanically noisy in addition to the exhaust noise. It was noisy the way your car engine is if you lift the hood and advance the throttle. This was not a sound suitable for a quiet house. Furthermore, the exhaust noise was still present just outside the house.

The generator was securely vented, but it still gave off fumes and smells in the cellar. No engine has perfect seals in the cylinders. Almost all of the fumes are forced out the exhaust pipe, but a small amount of gas passes by the piston rings, out into the crankcase, and then out the oil filler pipe. Consequently, when I ran the generator, a small amount of exhaust gas and a good amount of oil smell ended up in the cellar. Whenever I ran the generator, I really had to leave the cellar door open. This defeated the purpose of operating the generator in the cellar.

After the first winter, I moved the generator out of the cellar and under cover. I now had to go out the cellar door to start the generator. This worked moderately well until the day the generator broke down. I had to freeze while I repaired it. I soon realized that this 600 pound generator would become a liability at a time when it would need service above and beyond what I could do. I would have to drag it to a pickup truck, hoist it into the bed, and drive it to Hartford, Connecticut.

The negative aspects of living with the generator compelled me to advance my alternative energy system to a point where it could power my one large problem load, my deep-well pump. In a few years I managed to power the deep-well pump and all of my other 120 VAC house loads with an inverter and a solar electric system. I put the generator up for sale in the newspaper.

Somehow the generator got its last licks in anyway. The ad ran in the paper for weeks with no response. I was disappointed. I needed the $1500 for which it should sell. I was enrolled in a Masters program at Antioch New England Graduate School in Keene, New Hampshire. I returned late on a Friday night from Keene only to have the inverter blow up and cease to function. Luckily, I could still pump my water with the unsold generator. The next morning a man from the Tri-County Fairgrounds showed up unannounced, to see the welder-generator. He wanted the generator immediately. I needed the $1500. The generator happily left to its new incarnation as a traveling welder. I am sure it was happy to know that I lived without water and showers at home for three weeks while the inverter was being repaired.

One would think that life in the backwoods of New England where there are no power lines would be quiet and serene. Through the years I learned from my old customers that generators were the cause of many bad feelings between neighbors. A person in his or her own home can run a generator for a whole day and put up with the noise as a necessary evil balanced by the benefit of the service provided. The same person does not appreciate the noise of a neighbor's generator. Often the neighbor has a solar electric system that is too small. Then the generator runs every day all winter long. In the worst case, the neighbor is a weekender, who brings along his small portable generator to run all evening to power the TV and the lights in the little cabin.

If you drive down Bashan Hill Road past our home, you will soon intersect with Parish Road. If you turn left on Parish Road you will travel two miles on this dirt road before you reach Route 143. Half the length of the road wanders next to the state-protected upper end of the middle branch of the Westfield River. There are a few camps over the two miles. The first camp is about a mile from our house by road, about half a mile as the crow flies.

My wife Lea and I have little in common with the people in the camps. They come on weekends in the months of no snow or mud. We often see these people while running or bicycling. Since we do not know their names, we have had fun giving their places names. One camp is "Bending Lady Camp" because they have one of those painted outdoor plywood cutouts in their yard of an old lady bending over to weed her flowers while she exposes her wide thighs and polka-dot bloomers.

We have named our nearest Parish Road camp "Barking Dog Camp." Lea is very afraid of dogs. A retired world class runner, she fears an injury as the result of a dog bite. Dogs seem to sense this and give her appropriately bad receptions. This camp has three dogs that bark and threaten until their nice owners drag them into the camp.

The people who own "Barking Dog Camp" are polite and friendly. We would never know they were there if it were not for their generator. The first two years they owned the camp, they used their generator from dinner until they went to bed at eleven o'clock. They apparently used the generator to cook dinner and to run lights and a TV. They did not like the noise of the generator near the camp, so they placed it in the back of their pickup truck with the back door of the cap open. A hundred foot extension cord led down to the camp. They parked the truck with the open back pointed away from their camp and up toward us. I never complained to these people. Their generator was a half mile through the woods from us. When they were watching TV, it probably seemed noiseless to them. However, when we went to bed next to the upstairs window on a hot summer night, we could hear the whine of the generator very clearly one-half mile away. We much preferred the bird songs that we listened to when the weekend was over.

Our nearest camp is a third of a mile away down Bashan Hill Road. I have known Phil and Ann since I moved here; consequently, their camp has no funny name. Phil is quite clever. He has devised several unique systems for the needs of his camp. He has always had a generator as the center of these systems. He has always been conscious of the generator noise and has kept the generator in an outbuilding that retarded the sound.

Phil heated the mobile-home camp with a propane space heater. One winter Phil devised a system to heat his trailer using the heat lost from his water-cooled, propane marine generator. Instead of circulating the coolant to a car-type radiator, he circulated it to a home heating radiator in the camp. When Phil was there for the weekend, the generator heated the camp at the same time it provided electricity. This was not too bad for me, the neighbor down the road. His generator was not too loud; he was only there an occasional weekend, and the noise did not come into my house in the winter because my windows were shut. Phil tested the system out and found that he could use slightly less propane heating his house this way rather than using the propane heater. I doubted this, knowing the basic laws of physics, and knowing the cost of the limited life of the generator.

Phil later decided to heat his camp all winter long. That winter his generator was set to turn itself on whenever the camper needed heat. I listened to his generator anytime I was outside the house. I was patient. I waited for the inevitable failure of the overworked generator. The next winter the system had changed.

Phil owned a refrigeration wholesale parts warehouse. He had contacts for assorted equipment at wholesale prices. He also had a seafaring boat. He was very knowledgeable about marine power systems. Over the years I have watched him go through four heavy-duty generators. He maintained them well. Generators are like cars. They have many moving parts. These parts can and do fail. The generator runs unattended. When some part starts to fail, unlike a car, there is no operator to notice the beginning failure. The part completely fails, often causing additional damage to other parts or complete destruction of the engine.

Our neighbors, the Cooks, have always had a solar electric system less than half the size of our system. Bob is a welder. He needs a heavy-duty welder-generator. They have always used this generator to supplement their solar electric system. The generator is installed in the garage away from their house and nicely muffled. We certainly cannot hear it from our house. Lea and I do pay attention to the noise when we are visiting the Cooks. We can see that this would not be for us.

The noise of a generator can spoil the good part of a project or the fun of a system. The Cooks are very close to Lea and me in their lifestyle. Because of this similarity, and because of my natural analytical nature, I like to watch the progress of their homestead projects. Lea and I get to try a few things out before we purchase them.

Two years ago our business sold an Onan portable generator to Bob Cook. Bob needed this generator for a specific load he could not supply from his large welder-generator installed in his garage. The specific load was a set of fans in the wood burner of a maple sugar evaporator. The Cooks' maple sugar house is a few thousand feet away from their home and garage. Bob was upgrading his old equipment to a state-of-the-art burner that would greatly decrease the time it takes to boil. This new equipment allowed Bob and Karin to boil sap to syrup in significantly less time with considerably less labor.

Bob has spent years working at construction sites where a generator is constantly running all day, every day. I think he never even is aware of the noise. I have loved hanging around sugar-houses since I was a small child and hung around our sugarhouse when Fran Granger tapped our sugar bush. The quiet and the aroma are peaceful. Lea and I still visit the Cooks' sugarhouse, but it is no longer the same. The noise in the background is too much. If Lea and I were the operators of a sugarhouse we would have continued the old way. The peacefulness would have been one of our reasons for choosing that line of work over something else.

We sold a lot of Onan Generators while I owned Fowler Solar Electric Inc. We always tried to get people to buy solar electricity first. We liked to see the generator used only in a capacity of supplying backup power when the winter sun refused to shine during the bad cloudy spells of November and December. We liked to see people buy a less expensive, but good, portable generator instead of a heavier-duty stationary generator. The difference in cost could be put into the solar electric modules to produce more solar electricity. The lighter-duty generator would then be used less. A broken portable generator could be transported in the trunk of a car to the generator service center.

Lea and I have not had a generator since I sold that Miller welder-generator. We are purists. I have always told Lea that if we were customers of my old business, I would advise us to purchase a backup generator as the next logical expansion of our alternative energy system. So far we have had enough power that all we need to do is conserve in the winter.

CHAPTER 11

More Building the House

THE BIGGEST emotional scare of my house project came one hot August morning. It was noontime and Rolo had been gone for several hours. Rolo often disappeared for periods of time, chasing rabbits, or taking a trip to the Cooks' to mark out some territory for their dog Annie. He rarely left on an adventure for more than half an hour. Rolo was the type of dog who always kept a watchful eye on his master. The jingle of keys or the lacing of boots were signs that the master was about to go. He never wanted to be left behind.

Scott and I kept a constant lookout for Rolo during the course of the day. We watched to see that Rolo was not asleep at the base of the foundation before we threw scraps from the building, and we always located Rolo when we heard a delivery truck approach. Normally Rolo slept in the camper or under the camper to stay out of the hot sun.

This August day was a typical hot and sunny day. I had called Rolo several times over the past few hours, but he had not responded to my calls. We were engaged in some heavy work on an upper story; I had not thought to stop and search for him. When we broke for lunch, I realized that I had better find him right then. Something was wrong. Scott and I searched for Rolo in all the normal places. We checked other places he might have been hiding on this hot day. Rolo was not to be found, and, even more strangely, he was not coming to beg at lunchtime.

I tried to retrace the morning in my mind. I finally remembered that I had unloaded some supplies from the hatchback of my Dodge

Colt. I had problems that summer with Rolo jumping into the car whenever I opened a door. He was so ready to go for the next ride. This was a real danger because he could get shut in the car on a hot summer day. I was sure I had not opened any doors in the car, but I knew this was the next place to check.

Of course Rolo was in the car. Rolo was a very good jumper. I had opened the hatchback and carried some supplies to the cellar. Rolo wanted to go where I was going and had jumped the several feet, up and over the hatchback lip, into the front seat. By the time I found him at noon, the car was overheated enough from sitting in the sun to have killed a dog. Rolo was both lucky and smart. He was lucky that he was small, and he was smart enough to have crawled to the shaded floor under the dash. The upper part of the car was extremely hot, but the floor was less so.

Rolo was panting badly when I pulled him from the car. Scott got the garden hose, and we wet him to cool him off. In a few minutes he was fine. We had had a close call.

I had planned to insulate the basement of the house with two inches of closed-cell foam insulation. Normally I would have installed this insulation immediately after I had capped the foundation. I postponed the installation of the basement insulation because the backfilling was delayed. This type of insulation degrades in sunlight. I waited until I had word from Zack that he would soon backfill my foundation. When Zack finally arrived, the whole house was framed and sheathed. We finally insulated the foundation after the whole house was built.

The insulation of a foundation was relatively new and uncommon in the hilltowns. Every workman had a caveat and a hearsay story. I heard how the frost had pushed the insulation up with such force that the insulation buckled the siding off the wall. I was told that the right way was to insulate below grade, and not to insulate from the ground up to the siding. Other people were recommending that I insulate on the interior walls of the foundation. I also heard that carpenter ants loved to live in the insulation.

I tried to listen to and learn from the gossip on the job. I really gave some thought to all of the ideas and the worries. Some ideas, such as not insulating from the ground to the siding, were in error. I have in fact seen several houses insulated this way. The part of the foundation exposed above ground is the area of greatest heat loss from the basement. This is the place where the wall is exposed to the coldest air temperature. Underground, the insulation is less necessary because the earth is a moderate insulator and the winter temperature of the earth is normally warmer than the winter temperature of the air above it.

It sounded quite likely that the frost could push the insulation up and buckle the siding. My solution to this was a very solid baseboard below the siding that would be securely fastened to the box sill. I guessed that the insu-

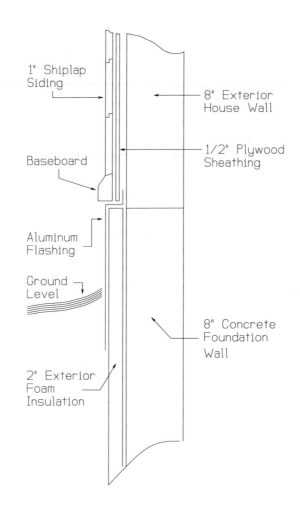

1" Shiplap Siding

8" Exterior House Wall

Baseboard

1/2" Plywood Sheathing

Aluminum Flashing

Ground Level

8" Concrete Foundation Wall

2" Exterior Foam Insulation

My method of covering the above-grade area of the foam insulation around the foundation.

lation could only move the lightweight clapboards or shingles on a house. I did not want the foam insulation on the interior walls of the cellar where it would off-gas into the cellar, then up to the living space above. Furthermore, the carpenter ant idea sounded possible. I thought that if it were true, the ants and I would be happier if they made a nest in the insulation outside the cellar, rather than inside the cellar.

My greatest concern was how to protect the aboveground portion of the insulation from degradation by sunlight. This was a situation where the installation process of a new construction material needed time to evolve. I found no great solutions in my library on solar construction. My solution was to cover the insulation with a sheet of aluminum flashing. Scott and I custom bent a 24-inch-wide sheet of aluminum, such that two inches fit against the box sill of the house, where it was nailed and secured. The first bend sent the aluminum horizontally out over the insulation, and the next bend sent the aluminum vertically down the face of the foam and below grade. Later a thick baseboard at the bottom of the siding was nailed over the aluminum on the box sill. This would hold the foam when and if the frost pushed it up the wall.

The greatest heat loss from a cellar is through the concrete wall above the level of the ground. An eight-inch concrete wall only has the insulating value of a one-inch thick layer of wood. To best use insulation, the greatest amount of the substance should be installed between the area of the warmest inside temperature and the coldest temperature outside. This is where the heat escapes the fastest. In this house the south facing slab would be the warmest place that the cellar wall would be exposed to the coldest outside winter air. I decided to double the insulation in this area from two inches to four inches. I had purchased enough foam to insulate the foundation walls from top to bottom. The recommended method was to insulate down to the maximum frost depth. I insulated the south, east, and west walls around the slab, from the top to the bottom of the foundation. On the other walls, I insulated four feet below grade, the maximum frost depth in this climate. I installed an additional two inches of foam insulation around the perimeter of the solar slab, and down two feet from the siding. Above the four inches of foam surrounding the slab, I increased the thickness of the baseboard such that it would prevent the full four inches of foam from heaving from the frost.

Today, most foundations in the Northeast are insulated with foam on the outside of the walls. The new concept has developed into an accepted building method. Other products and building procedures have evolved to aid the new method. Currently, the foundation of the house is undersized two inches on all walls to accommodate the two-inch layer of foam insulation. The house is framed such that it overhangs the foundation by two inches in all directions. The outside walls of the house are in the same planes as the outside surfaces of the foam insulation below them. The house sits in the same orientation to the foam insulation as it would to the foundation wall of a house that has no insulation. The foam insulation is protected from sunlight by a layer of special stucco that is troweled on to make it look like concrete.

The exterior walls on my house were to be framed differently than if I had used the normal 2x4 or 2x6 framing construction methods. These walls were a double 2x4 construction. The 2x4s in each wall were 24 inches on center, but the studs in the inner wall were out of phase with the 2x4s in the outer wall. If you followed the studs down one of these walls you would see a stud in the inner wall, then 12 inches later a stud in the outer wall, then 12 inches farther a stud in the inner wall. No stud ever went any deeper in the wall than halfway. Later, when I added the insulation, any stud was fronted or backed by four inches of fiberglass insulation. This quilted effect made the wall better insulated than an eight-inch wall constructed of 2x8 studs.

This double four-inch wall construction has been used for years in commercial buildings along selected walls where sound transmission is a problem, such as a wall between an apartment and a lobby. I knew how to frame a 2x4 wall. Unfortunately, the solar articles that recommended this double wall construction did not detail the framing methods. Before I built the house, I tried to visualize all the places in the framing of the house where this double-wall framing would differ from conventional framing. Later, I continually faced new situations as they arrived on the job. I think each corner of the downstairs walls is framed slightly differently.

My first decision was to use an eight-inch plate on the top and the bottom of the wall, rather than two different four-inch plates. This meant I could frame the double wall in one unit, stand it up, and put it into place instead of building and installing two four-inch walls. On top of the top plate, I added the customary second, or double, plate. This too was a 2x8.

Ordering the windows required more decisions. A conventional window at the lumberyard is dimensioned to be installed in a wall composed of 2x4s in the middle, half-inch plywood on the outside, and half-inch Sheetrock on the inside. Windows for 2x6 walls are sometimes stocked and sometimes available only by special order. My windows would be very special order because they would have to fit an eight-inch wall that was the full eight inches of my full-dimension rough-cut lumber. I chose not to pay the exorbitant price of special order windows. I ordered standard windows for a four-inch wall. I would custom build my own jamb extensions to make them fit the full depth of the wall.

The imprecision of the unplaned lumber from the sawmill was a constant headache. The 2x4s were especially irregular. I found that my double-wall construction was very forgiving of these badly dimensioned 2x4s. When I used eight-inch plates and flushed the outer wall 2x4s to the outside of the plates and the inner wall 2x4s to the inside of the plates, all the too-wide and the too-narrow of the 2x4s were left in the middle of the wall, where it would never matter. The 2x8s were also irregular. We had to sort them for width so that the wall made a gradual transition as it changed from seven-and-three quarter-inches wide to eight-and-one-quarter inches wide. The most difficult variation in size was the variation of up to one-half inch in thickness of a 2x8 plate. The walls could vary a half inch in height because the plate at the bottom was irregular. I could call it farmhouse construction. The real problem came from the fact that there was the thickness of first the bottom plate, then the top plate, and then the second top plate. In the worst scenario, all three plates could be one half inch too thick and could triple the error one-and-one-half inches. We constantly sorted and fudged to keep the wall height consistent.

The 2x8 and 2x10 joists also varied up to half an inch. Bob Bartlett had taught me a trick several years before. I sorted the joists before I laid them out. Normally the joists were consistently inconsistent, in that a series of 2x8s would all be seven-and-three-quarters inches at one end and eight-and-one-eighth inches at the other end. I sorted all the fat ends in one direction and all th thin ends in the other direction. The small change from the one side of a room to the other was imperceptible, while the wavelike motion of thick then thin joists next to each other would have been very obvious. I remembered that historically, sorting by size had been a standard part of carpentry. It was only in recent history that framing lumber was dimensioned.

WHEN ZACK ARRIVED to backfill the foundation, we had framed the whole house and were finishing the eaves in preparation for shingling the roof. Zack was very apologetic when he saw us working on the eaves from ladders that stretched up from the base of the foundation instead of from the ground. He said he would have come earlier if he had understood. I had forgotten to be a squeaky wheel.

The second day Zack was on the job, we were interrupted from our work by a strange and unsettling visitor. Down the driveway and into the path of Zack and his 'dozer, drove an old beat-up station wagon. The driver abandoned his car and motioned rudely to me to come down from the eaves. I ignored his early morning beer breath and tried to be polite. His first words were, "You're building this house on my land!" I was so caught off guard that I inwardly panicked. I looked around at my house that was framed, at the well that was dug, and at the septic system that was almost complete, and imagined moving them because I had been so stupid as to buy a piece of land that was not deeded properly. I thought longer, panicked further, and called my lawyer.

My patient lawyer explained that the deed was properly researched. There was no mistake. He further pointed out that the title was insured anyway. He asked if I had measured my boundaries correctly. There was no question about my measurement of the boundaries. The house was in the middle of the nine acres. The starting point for all measurements was a historic stone tri-town marker. I now realized where the error was. This jerk did not know where his land was. I asked him for the survey number of his lot. I got out my survey map of the registered subdivision and explained to him that he was in Worthington, and that his lot was down the road a thousand feet in the town of Peru. After I refused a beer and told him I did not have time to visit, I went back up the ladder. Zack had heard the whole thing and told the man his car was in the way, which finally got him to leave the premises.

ONE SUMMER EVENING I drove down to Clarence's sawmill to pick up some lumber. Buying lumber from Clarence was a little more low key than going to your average large lumberyard. Clarence's sawmill was up the road

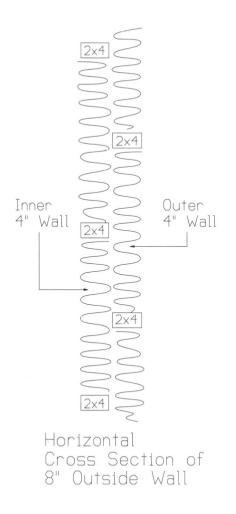

Inner 4" Wall

Outer 4" Wall

Horizontal Cross Section of 8" Outside Wall

Horizontal cross section of the studs and insulation in a double wall.

and across the street from his house. About two-hundred feet from the road was a large gate that could be closed and secured with a chain and lock. Personally, I have never seen it closed or locked. Next to the mill, 300 feet beyond the gate, is a giant rack that holds assorted framing lumber. Clarence instructed me, as he had instructed most people, to go in at any time and load up. If no one was around, I should leave a list on the door of his house and pay him later.

I had loaded my little hatchback car with 2x4s until the bumper nearly touched the ground. Before I got into the car, Charlie Nugent and a friend drove in. Charlie introduced me to his old friend Charlie Anthony. Charlie Anthony had grown up in Chesterfield, but he had been in Florida for many years working construction. As the conversation wandered, I learned that Charlie Anthony was a mason. He would be around for the summer and needed some work. Before they left I made an appointment for Charlie Anthony to stop over at my house and give me an estimate on my chimney. Later, when I stopped by at Clarence's house, I got a good recommendation of Anthony's masonry abilities.

I needed a double, eight-inch flue-lined, standard block chimney, about 32 feet high from basement floor to the peak of the roof. One flue was designated for the Vermont Castings Vigilant combination fireplace and efficient downdraft stove that I planned to install in the living room. I had had the same model stove in my previous house. In the living room, I would later stone face the only exposed ugly chimney block. The rest of the chimney block was to be either hidden inside walls or closets, or stuccoed and painted to match the wall in the hallways.

As the stove heated, the mass of the chimney would be heated and would radiate heat into the center of the house rather than to the outside. The second flue would be used only for venting the hot-water heater in the basement. A lined eight-inch flue would be excessive for a small gas appliance. In this case I looked at the second flue as a backup flue. If the wood stove flue ever got damaged from age or a chimney fire, I could always switch over to the second flue. At that point the gas hot-water heater could be switched to the damaged and repaired flue. I would also have a small wood stove in the basement for the times when I would work down there. The basement wood stove would legally vent into the same flue as the upstairs Vermont Castings wood stove, whereas the building code prohibited the venting of a gas appliance into the same flue used by a wood stove.

Charlie Anthony showed up for his evening appointment. There was no problem with the straightforward job. Charlie found all of my framing to be accurate and plumb. He would require no changes to run the chimney up through the existing holes. The spaces for clearances between the wood of the house and the concrete block were up to the latest and recently upgraded code. This was the deal: I would be Charlie's grunt. I would mix the mortar and carry the chimney materials up the ladder to Charlie while he built the chimney. I would buy all of the supplies on my account at the best price I could get. I would pay Charlie $175 for labor. After I purchased the materials and paid Charlie, I had a double flue-lined chimney for only $600.

Being the grunt for a chimney mason is a very hard job. The chimney starts in the basement. The grunt has to carry the sand and bags of mortar mix down to the cellar. Most part-time masons mix the mortar in a wheelbarrow-like container with a hoe. They use a cement mixer only if it is a much larger job. The mortar must be pretty thick. This means a lot of hoe work. The grunt also has to carry all of the chimney block and flue tiles down to the basement. As the chimney goes up through the house, the grunt has to start carrying all the same things up the ladders to the upper floors, and eventually up the ladder and up the roof. The already heavy blocks, tile, and buckets of mortar get heavier as the grunt awkwardly makes his way up the ladders.

The first day that Charlie and I worked on the chimney was the day after my trip to the Worthington Health Center for carbon-monoxide poisoning. Scott was not working. It was a hard day, but I completely flushed the poisons from my system. We built 80 percent of the chimney the first day. The second day we finished the chimney and installed the flashing.

I feel I made a mistake on this job that I made several times during the house project. I talked too much to the man I hired. If I had been the mason, I would have preferred a grunt who was quiet, so that I could concentrate on my work. I often made Scott stop talking about non-job topics when I needed to concentrate. However, when I was the grunt, I talked a lot. I wanted to know the person and make him comfortable. On this job, Charlie forgot to place galvanized ties between the chimney block where I would later use them to secure the stone facing in the living room. If he had had a silent helper he would have remembered. It caused only a little extra work later.

I found one major problem with the chimney in the fall when I first started using the wood stove. The stove would backdraft smoke into the room whenever the wind gusted. The chimney exited the roof seven feet north of the peak; it stood seven feet above the roof, level with the ridgepole of the house. If the chimney had exited the roof at the peak Charlie would have built the chimney two feet above the peak. In this case he felt the height was adequate given the seven-foot horizontal distance of the top of the chimney from the peak of the roof.

I called Vermont Castings about the problem. My first worry was that the eight-inch flue was too small for this stove. If that were true, I would need to change the stove. It was certainly too late to change the flue size of the chimney. Vermont Castings clearly stated that the backdrafting was a result of poor draft caused by a chimney that did not rise a minimum of two feet above the peak of the roof. I could not believe that two feet could make such a difference. I next went to the library. All the reference material stated the same thing. The chimney must be two feet above the peak if the peak is closer than ten feet from the chimney.

I had a hard time understanding that this short change in height would make a difference when the chimney was only three feet short of the required ten feet from the peak. I questioned some local workmen. They felt I should extend the chimney. I finally took an old piece of aluminum flashing and fabricated it into a square tube to match the chimney flue. I inserted this into the top of the chimney to make a crude and temporary, two-foot extension to the chimney. The problem was solved. The old maxim of two feet above the peak if the chimney is closer than ten feet from the peak was really a very precise empirical statement of aerodynamics. I removed my test extension. I used my mason's experience from working with Charlie and extended the chimney permanently.

One morning in late July, Scott arrived at work to announce that he would be leaving in two weeks. This would be three weeks earlier than we had planned. He decided to go on an orientation backpacking trip for college. I had never heard of this before. I had tried to be very explicit in the spring, when I had worked out our work agreement, that I would need him until September. Because Scott had taken some time off, and because he had been sick, he had only averaged 30 hours per week. I was very disappointed. Scott was now trained on this job. I did not have money to hire a carpenter.

I tried to remember how I had canceled my summer job at the general store after my sophomore year of college. I just couldn't bear to return home to Worthington for the summer. I knew I was leaving my employer and best friend, Pete Packard, in a bad position. I had only been capable of making the best decision for me. Scott was in a similar situation. I tried to be as understanding as Pete Packard had been to me. I resigned myself to a tough three weeks in August. After all, three weeks was not that long.

My only choice was to hire Willie Brown, Scott's eighteen-year-old friend. I had literally known Willie since he was born. Willie was an intelligent young man, but he was untrained in carpentry. Willie worked differently than Scott. He always arrived late. On the other hand, he never watched the clock at the end of the day. After a week I learned to note what time he had arrived, and to not announce quitting time until eight hours later. He never noticed he was a half-hour late arriving, or that I kept him a half-hour late. Willie had a good attitude. He was slower and more deliberate than Scott, but also less likely to make an error. He had two things in common with Scott: he loved being at the site, and he really felt a part of the project.

CHAPTER 12

September Fifteenth and Cold Nights

ROM MAY 18, when I took possession of the land, I had one major objective: I wanted to have the house framed and closed in by September 15. The nights would be cold every night by that point in the higher altitude of Worthington. I did not want to return to the frosty mornings of the camper and the tent. I worked long hours and weekends to meet this goal. Whenever Scott or Willie could not work, I just worked their hours in addition to my regular hours and kept on schedule.

As September 15 approached, the house was nearly closed in. It was framed, roofed, and sheathed. The windows were also in place. The only remaining projects were the installations of the front and the rear doors. I had bought used doors for which I had to make custom jambs and thresholds to fit my extra thick heavily insulated walls. Jeffrey Hartman came out from Boston on the last two days to assist me. We ripped, planed, and routed the lumber. On exactly the fifteenth of September, the doors were in place. Jeffrey left for Boston, and I left for Mrs. Joslyn's barn.

I had stored my furniture in the upstairs of Mrs. Joslyn's old barn in town. I chose one soft rocking chair, kitchen chairs and table, a wooden rocking chair, an old couch, and a single bed mattress. The VW camper was full. I went home and swept up the plywood sub-floor of the living room. I arranged my first furniture in the new house. On the north wall I piled about a hundred 2x4s in stacks about eighteen inches high, in the approximate dimensions of a single bed mattress. I placed the mattress on the top for my bed. The living room became my little studio apartment amidst the construction materials that occupied the remainder of the house.

I still had no toilet and only cold water in the basement. However, at this point my life returned to sanity from the four-month push of the summer. I took my first real time off and drove to Boston to visit friends. It was good to get away from the grind of long days with no breaks. In Cambridge I purchased the Vermont Castings wood stove around which the living room was designed. We loaded the 350 pound stove into the hatchback of my poor Dodge Colt, and Rolo and I headed back home.

I sold the VW camper to a friend who had been waiting for it to become available and gave the old tent to Willie. I now lived in the house. I tried to keep my living space as clean as possible. Parts of the house that were under construction were normally isolated by polyethylene walls. Unfortunately, the majority of the house was filled with the lumber for the outside siding and the inside flooring and trim. I was happy to be warm and toasty with my new wood stove.

My work schedule backed off to eight or nine hours a day with time off on the weekends for social events. I began to feel like a real person again. Many people who build their own houses back off so much at this point that they become stalled and unproductive. I maintained an attitude that this project was a job. I made myself start work on the house on time at eight a.m., break for lunch, and work until the end of the day. Still, I kept each project well planned and stocked with the necessary materials and tools to prevent delays.

IT WAS A WARM FALL, and Willie was available for work. I had planned to leave the outside of the house unfinished. I would cover the plywood sheathing with construction paper until spring. Willie worked part time when the weather permitted, and we gradually sided the house. The outside work was a welcome balance to the inside work. I also enjoyed having someone to work with. It was lonesome working inside all day by myself.

My first inside work was finishing the interior framing. There were many places where we had used short cuts to keep the big project of closing in the house on schedule. I now had to finish the joists that were left out, frame the holes for the stairways, and add the interior walls. My priority was to complete the framing that was necessary for the plumber to plumb the house.

Darrell Shedd was my plumber. He was, first of all, the plumbing inspector in town and, second of all, my neighbor and friend. It is always wise to hire the workman who is also the local inspector. The job is performed to the most recent standards and easily passes by the alternate inspector, who inspects the local inspector's work. Darrell was also an excellent plumber. He and his dad owned a large plumbing company in the nearby city of Pittsfield. Darrell still liked to do some side work close to home, to balance his more foreman-type work at the company.

Darrell was a perfectionist. He wanted his plumbing to not only work well but also look good. He refused to use less than the best materials. He refused to install the cheap toilet I had purchased. There was no compromise, I returned the toilet to the supplier for a better model. Darrell was correct; it was not worth installing.

I had designed the plumbing and the placement of the kitchen and the bathroom such that all of the plumbing was on the first floor. This was a simple, straightforward job. I asked Darrell to plumb the supply pipes such that they were always continuing uphill to the fixtures with no dips or valleys. This meant that they also went straight downhill back to the tank in the basement. When I would want to drain the pipes, the drain could be opened in the basement, and all of the water would flow back to the basement. Draining the pipes would be a safe way to leave the house for extended periods of time in the winter if my sizing of the passive solar heat was in error. Darrell always liked to plumb with this method.

Early in the summer, I had stopped at my old friend Lee Monroe's house in Chesterfield. Lee had been Sue and Kenny's best friend and neighbor in Worthington, and consequently my good friend. He and his family had moved to a house next door to Bisbee Bros. Lumber Company. Lee and I had done many favors for each other over the years. He had just upgraded his bathroom. He knew my tastes and offered me a claw foot tub in perfect condition and the old sink to match. I was delighted. Lee even delivered them to me the next day with his trailer.

From a bulletin board advertisement at the general store, I got the telephone number of someone who was selling a gas hot-water heater for $100. I called and found the person selling the hot-water heater to be my old friend Greg Donovan, Zack Donovan's youngest child. The heater was nearly new. I bought it. Later in the summer I bought a gas stove from another ad on the store bulletin board. This stove was an old Glenwood gas range with a space heater on one end. The heater would be a safe source of backup heat in the kitchen if my solar heating was inadequate. Ultimately, I never hooked up the flue because the heater was not necessary. A few years later I replaced the Glenwood stove with a smaller, conventionally-sized stove.

DARRELL ROUGHED OUT all of the plumbing and ran the gas lines to the water heater and the stove. The kitchen counter and sink would not be built for many months. The bottom four feet of two of the bathroom walls would be mortared stone walls. Since the bathroom sink would hang on one of these walls it would also be plumbed much later. I longingly anticipated the installation of the bathtub and the hot-water heater.

The whole front half of the downstairs of the house had no structural interior walls. Eventually there would be walls around an area that would be the bathroom. At this time the downstairs was one open area from the living room, through the future bathroom, and into the kitchen. Out in the middle of this 14x36 foot area Darrell and I installed the old claw foot bathtub and a toilet. The south wall of the 36 foot area was all windows.

Darrell had capped the pipes that would feed the future kitchen and bathroom sinks. One Wednesday morning Darrell announced that he would complete the toilet and the bathtub installations that day. "You'll be able to have a hot bath tonight," he said. That statement made my day. The Cooks' pond was too cold, the cold water in the basement was oppressive, and I was tired of asking friends for showers. All day long I anticipated my Bashan Hill baptism. At four o'clock Darrell announced that the cold water was working, but the hot water would not be completed that day. He had to leave early for a Board of Health meeting.

Darrell never mentioned my hot bath before he left. He certainly could not have know how much I was looking forward to my first bath at home in five months. Darrell did not return for two weeks. I just kept washing in the cold water. Finally Darrell returned to connect the hot water. As soon as he left, I dove into my hot bath even though it was only three o'clock, long before quitting time. It was a holiday: the coming of the hot water.

Many processes that are automatic in a conventional house become quite manual in an alternative energy house. In an unfinished alternative energy house, these processes become even worse. For example, I had to go to the cellar to start the generator before I could take a bath. After the tub was filled, I then had to return to the cellar and turn off the generator in order to save fuel and to enjoy a quiet bath. After a few baths I had perfected my system. I would run the generator until the bath was filled, then shut the generator off. The dirty dishes were sitting in a plastic tub next to the bathtub. After my bath, I would place the plastic tub in the bathtub and wash the dirty dishes with dish soap in the bath water. The generator had run long enough after filling the tub to also fill the large water tank in the basement. I would drain the tub water turned dishwater, rinse the dishes, and place them in a dish rack in the now drained bathtub. At the end of this cycle, there would be plenty of water left in the water tank to flush the toilet and wash up for the next twenty-four hours until my next bath and generator run.

If I operated power tools at any time during the day, the generator would concurrently power the water pump, which would fill the water tank. This large tank held 120 gallons of water. It had a pressure switch that would turn the pump on when the pressure dropped below 30 psi (pounds per square inch) and shut the pump off at 50 psi. In a grid house the tank would yield about 30 gallons of water to the first floor before the pressure would drop to 30 psi and turn the pump on. In my house the pump would turn on at 30 psi only if I had the generator running. Without running the generator, I could get an additional 30 gallons as the pressure in the water tank diminished to 5 psi. If I did not use the generator during the day to power tools, I would run the generator specifically to power the pump. Six minutes of generator run time pumped the system from 5 psi to 50 psi.

Toward the end of the summer, I had called Bill Wilson, the electrical inspector, to see if he would like to do the wiring in my house. I did not know Bill at the time, but he was recommended to me by several people. Bill answered gruffly that he was so far behind and had too much work to do, so I quickly took another tack to keep from further offending him. I asked if he could recommend anyone else. In his typical, unpredictable way he asked me why I did not want him to do the wiring. I had accidentally taken the right approach. Bill could not bear to hear of someone else doing the wiring, because he did not trust any other electrician to do as good a job as he would. Bill was a true master electrician.

Bill arrived on the job soon after Darrell had finished the plumbing. Bill was about 70 years old and as cautious as they come. He stared at the inside of the house, and he stared at me, his face shadowed by a pair of big black eyebrows. His hands looked strong enough to squeeze 2x4s into sawdust. These hands had spent a lifetime twisting stiff wires into small places where they did not want to go.

Bill laid out the boxes in the walls and the ceilings, making big black marks with a marker. He asked me questions about rooms and appliances as he worked. Bill left me with some homework to do before his next visit. There were walls that needed to be finished and some framing that needed to be changed. This was the first in a long series of homework assignments from Bill.

The next time Bill came to work, he was an hour late. He started the conversation with his patent apology, "I got loused up." He then filled me in on the latest crisis of some little old lady whose electric stove was not working. He could not leave her inconvenienced, so he had gone to her house first. He knew what parts he needed for the lady's stove and would be leaving my house earlier than planned to go to Springfield for them.

Bill started to warm up that day. He remembered me and my friend Paul Dunlevy from when we were in high school. Bill lived down the road from the landfill. Paul worked for old Emerson Davis, the caretaker of the landfill. Bill used to do some work for Emerson with his old 'dozer. It all started to click who this Mr. Wilson was.

Worthington was — and still is — too small to have its own high school. My sister went to Williamsburg High School, twelve miles away. The Town of Worthington paid tuition to the Town of Williamsburg for the students who attended. When I was a freshman I attended Williamsburg High School as my sister had, but the following year, the Town of Worthington sent its students to the brand new Gateway Regional High School in Huntington. Ironically, this regional high school was the successor to Huntington High School, from which my father was graduated. I often met

Bill Wilson
the electrician.

my friends' older relatives who had known my grandparents John and Irene Fowler, whom I had never met.

Paul Dunlevy and I were very good high school athletes. We had been unfairly blackballed from junior varsity sports our freshman year at Williamsburg High School because the varsity coach did not want our skills developed. He knew we would be playing for Gateway Regional the next year against his Williamsburg team. This early lesson made Paul and me more competitive. We took great pleasure in our junior and senior years beating Williamsburg every basketball game on our way to two consecutive league championships.

There was no after-school sports transportation from the regional high school to the member towns. Many kids from the distant towns did not participate in after school sports. Each day after practice, Paul and I walked half a mile from the high school to Route 112 and hitchhiked fifteen miles home. This was not too bad during the fall soccer season. However, after basketball practice, it was cold, dark, and sometimes snowing. One practice I suffered a severe contusion to my knee joint that put me out for the rest of the season. Paul helped me limp to Route 112 and then hitchhike home.

Bill Wilson was one of a few people who regularly traveled that road from Huntington to Worthington. He drove home much later than our practice. He was our backup ride on bad hitchhiking days. Bill would always stop for us and somehow fit us into his old battered station wagon stuffed with electrical supplies. It looked like the classic abandoned station wagon in the back pasture filled with old car parts, only this station wagon commuted forty miles to Springfield each day.

When Bill wired my house, he no longer had the old station wagon. He had a Ford Econoline van. The van was as over-loaded as the station wagon had been. My least favorite job for Bill when he wired my house, and for the next several years when I worked for him, was to be sent to find something in that van. He had almost any part we could possibly need, but it was hell to crawl through the piles of boxes, some spilled, to find the right piece.

Bill started testing me during our next work session. He asked me questions about controversial subjects such as Vietnam and then proceeded to argue with me. This was his way of building a friendship. Once he became sure of me, he became one of the most devoted friends I have ever had. Even today Bill is on the top of the list of people I can call if I am desperate for help of any kind.

Bill's pattern was to leave early for Springfield and leave me with some homework for his next visit. He would leave a heavy duty drill and bit for me to drill the holes for the wiring. Later he instructed me to run wires through the walls. He trained me to wire all the boxes. When I was sufficiently skilled in all facets of home wiring, Bill started hiring me on his jobs.

The first week I worked with Bill at my house, I wondered what would happen if he suddenly died in the middle of the job. He was having some health problems and often left early to go to the V.A. Hospital in Springfield. He looked strong, but he talked as if he were sick but too stubborn to lie down. I soon learned he was the toughest old guy I had ever worked with. Bill is just as tough today, and he's still working at the age of eighty-four.

Bill's first inclination was to wire a house to the best of his abilities. He preferred two switches in each room so one could travel throughout the house in light. One switch turned the light on in the next room as the switch next to it turned the light off in the previous room. He liked to install the maximum number of outlets in the kitchen with only two receptacle boxes per 20 amp circuit. This way high-wattage appliances such as toaster ovens or coffee makers could run on any outlet. Bill also wanted to place outside lights all around the house with additional switches in the master bedroom in case security lighting was needed in the middle of the night.

I wanted a minimal amount of electrical wiring in my house. I had decided early in the project that I did want the house wired up to code before the walls were closed in. I did not want an alternative energy home that would need to be torn apart and wired to code if the power line came down Bashan Hill Road at some time in the future. On the other hand, I was someone who walked to the bathroom in the middle of the night with no lights. I had grown up in an old house that had few switches and outlets. It would be no hardship for me to walk across a dark room to turn on a light switch. Bill and I bargained on many things. He would insist on multiple switches every-where, and I would counter that I did not need a switch at both the top and the bottom of the stairway. We kept each other in check.

The final wiring for my house was a compromise between a Bill Wilson state-of-the-art job and the nineteen-fifties job I might have wanted. We had a 120V service panel with room for expansion. There was one light switch for a central light in each room. The kitchen, bathroom, and work areas had plenty of 20 amp outlets with a minimum number of boxes per circuit. Hallways and stairways had switches at each end. All the rooms had plenty of standard out-lets to avoid any need for extension cords. The cellar had plenty of lights and outlets. Ultimately, I would use a small number of kilowatt-hours per month, and there would be few appliances or lights on at any given time, but the wiring would be everywhere, for convenience and efficiency.

At the beginning of the project, I had planned to have a wind generator in the near future. I would possibly install a 120VDC wind generator to charge a 120V battery bank and convert my standard 120VAC appliances to 120VDC. Circuit breakers in standard service panels cannot handle DC elec-tricity. However, the fuses in the old-style boxes would work fine on 120VAC current or 120VDC current if I installed a 120VDC wind machine. Bill and I waited for an opportunity to remove a clean, healthy 100 amp fuse box from a house in which we were installing a modern circuit breaker service panel. When we found the fuse box, we installed it at my house.

My potential wind system was never purchased or installed. I found solar electricity. I installed my first solar electric system before the interior walls were closed in. This system ran on 12 volts only. I ran additional wiring such that I had one 12V box in each room of the house. Today I still have these circuits, though they have been upgraded to 24 volts. Most of these circuits now run only efficient lights.

Bill was a master of the electrical code. Unlike many small town electrical inspectors, he had done heavy commercial work. While I worked for him, he got into many standoffs with electricians whose work he was inspecting. When the final decision was handed down, Bill was always right. I found it very hard to learn the electrical code from Bill. For our work, there was the Massachusetts Electrical Code and the more stringent Bill Wilson Electrical Code. Bill would never make it clear which was which. Whenever I pressed him to admit that he was requiring us to do something that was better than code and not legally required, he would put me off saying, "Do you want it to look like some goddamn shoemaker done it?" That was the end of the discussion.

Bill was very flexible about the electrical system for my house. We wired the house for smoke detectors in case the house was ever connected to the grid in the future, but we left the wires in the boxes and installed battery powered detectors next to the boxes. Ground-fault circuit breakers would not work in my house at the time. Bill let this requirement pass as unreasonable for this specific installation. When I eventually installed a sizable solar electrical system, he was very understanding that the electrical code had not yet been standardized for a solar electric installation and needed some custom interpretation.

The only real problem I had with Bill was the same problem that other friends had with him. It was murder getting a bill from him. I finally was able to pay him two years after he had started the job. He gave me an incredibly detailed bill of every screw he had used and a list of his one to three hour visits. My cash flow was very glad that the bill arrived two years after the work had been completed.

CHAPTER 13

Money and Mortgages

MONEY IS A LARGE CONSIDERATION on any building project. When I built this house, I was not a professional person with a strong and continuous employment record. I had had many jobs. I worked to be able to follow my interests in life and worried little about a profession for the sake of security. When I was a musician, I worked as a cook in a restaurant and later as an assistant manager in a vintage instrument store to pay the rent. I lived off of a part-time income repairing musical instruments from the Music Emporium in Cambridge when I returned to Worthington, where there were few jobs.

My financial position was such that no bank would consider me for a mortgage. I had no credit history. I had never had a loan or a credit card. In my lifestyle, I earned and saved the money before I spent it. My house project would have to be completed with cash and labor. I would proceed as far as possible and when I ran out of money, I would have to earn more before I could buy the materials for the next stage. There were advantages. I would have no monthly mortgage payments. Eventually, I would own a finished house. Sometime in the future, when I could again work a regular paying job, I would have money to save in the absence of rent payments or mortgage payments. I would eventually catch up financially to my contemporaries who had regular professions and mortgaged homes. For now, I would be equity rich and cash poor.

After buying the land for $13,000, I had $28,000 in cash with which to build the house. This was not an amount that would allow

me to have my house built by a contractor, nor was it even enough cash for me to buy all the standard materials necessary for a house of this size. I had enough money to acquire most of the materials for this house if I was creative in my choices, using less expensive alternatives such as native, green framing lumber, stone from my land, white pine floors, and used doors. Today, Lea and I look at our house and sometimes think it would have been nice to have spent $1,000 more on some specific material and have had fancier flooring or siding or paneling. I quickly realize that, if I had done just a few of these things, I would not have had the house at all. Our friends Sue and Chris Dunham live in the big old Colonial house where my childhood friend Jimmy Stevens grew up. The house needed absolutely everything when they bought it. They were happy it needed so much work, otherwise it would not have been for a price that they could have afforded.

When I was 30 years old, I worked at the Music Emporium in Cambridge, Massachusetts. This was a new business. Everybody who was a folk musician secretly wanted to work there because it was a special store, run by special people, and stocked with instruments that made the average folk musician drool. I was in the proverbial right place at the right time, and I was hired. My pay started at minimum wage and advanced only a little higher as my responsibili-

Our living room features an oil-finished, waxed, white-pine floor and a step-down to a Vermont slate floor in the south-side sun-space.

ties grew. No one was well paid. The store was not that profitable. I bought old instruments from flea markets, fixed them, and sold them on consignment to supplement my income. I lived in a house with four other people. My expenses were minimal. After a year I had several thousand dollars saved for a move back to the country.

My father died during my last summer at the Music Emporium. I inherited my father's mobile-home trailer and a few acres of land. I improved the trailer and grounds, and I completed the unfinished house I purchased from my sister. I sold both the trailer and the house in their improved states. I also sold musical instruments, wood stoves, a freezer, a refrigerator, and all other saleable items I could possibly live without. I had $28,000 in cash, the nine-acre piece of land on Bashan Hill Road, a one-year-old little car, and Rolo.

In many ways I felt very rich. I had more money than I had ever had in my whole life. I could order materials or labor, and just rip a check out of the checkbook to pay for it. Correspondingly, I knew this was not enough money. The well, septic system, and foundation cost $8500. I purchased a $2000 generator. This left only $17,500. From this small amount of money I needed to buy materials for the rest of the house, feed and clothe myself, and pay taxes and car insurance until I could find time to work again.

I designed this house to be a passive solar home with extra windows and extra insulation; at the same time I designed it so they would cost less than they would on a more common house. I shopped for price on materials and labor. I chose less expensive rough-cut lumber for framing. I made a bulk purchase of pine from Maine through Cummington Supply.

Cummington Supply was selling kiln-dried pine from Robbins Mill in Maine for 45 cents per board foot. This was two thirds the going price for other kiln-dried pine. The pine did have some liabilities. First of all, it was standard-grade, one grade lower than the normal lowest grade. This meant that there were bad knots to work around. The pine was planed only on one side and only shiplapped instead of the standard tongue and groove. Forty-five cents a board foot made up for the extra labor I would use to work with this pine.

I planned to use the pine for flooring and wainscotting inside of the house. The pine was planed only on one side because it was meant to be installed as siding with the rough side out. I considered using this as siding on this house. It would be much cheaper siding than $1.20-per-square-foot cedar clapboard siding, and much more durable than the cheaper alternative of pine clapboard siding. I decided to buy Cummington Supply's inexpensive pine and side the house horizontally with the shiplaps in a downward direction to shed water like clapboards. This siding would give the saltbox the look of an old mill.

I negotiated with Cummington Supply for a price on 5,000 board feet of 1x10 and 1x8 pine. This would be more than enough lumber for my siding, finish flooring, and paneling. I would rip the shiplaps off the surplus pine and use it for all the other jobs from eaves work to window casings. Peter and Fran

My wife's office in an upstairs bedroom paneled in 1x6 tongue-and-groove spruce.

sold me the 5,000 board feet of pine for 36 cents per board foot. When I started using the pine, I found it had more liabilities than I had anticipated; possibly I had not gotten the best batch. Many boards in this shipment were only partially planed on the planed side. I sorted all of the boards. The poorer ones went to the outside of the house where the rough side was out and the imperfect planing was hidden. I cut out the bad knots and used the shorter pieces where a shorter piece was needed.

By the time I was ready to finish the upstairs of the house, I was looking for alternatives to the use of wall board. I hated the dust associated with its installation which invaded the finished rooms of my house from the room under construction. I wanted to panel the walls and the ceilings in the bedrooms with wood. I started shopping around for tongue-and-groove pine. I finally found 1800 board feet of leftover 1x6 tongue and groove spruce. This was customarily used for ceilings in finished porches. It looked very similar to pine. No one has ever detected the difference in my house. Because this 1800 board feet was an odd lot at Northampton Lumber, they dumped it for 35 cents a board foot. Another good deal for my bargain house.

I bought doors from a friend who owned Rumplestiltskin, a recycled building materials store. I got eleven matching old interior doors and brass doorknobs. The doors had been removed from the nunnery at old St. Michael's School in Northampton before it was demolished. Ironically, the nice brass doorknobs alone were worth more than the $20 per door that I paid. The doors exemplify the way I saved money. They were better quality doors than the standard new doors. They matched the interior design of the house. These doors would have been a costly undertaking if I had had to hire a carpenter to perform the extra labor of making jambs and hanging the used doors. Since my labor had no cost, they were a better deal.

As the project progressed, there were many specific jobs I had planned to hire out, but later chose to do myself in order to save money. I had planned to hire Bob Bartlett to build a nice maple stairway. Bob was a better carpenter than I was, and I had never built a stairway. When the time came, I got a good price on yellow-pine treads, and I used short pieces from my other white pine for the risers. I carefully designed the stairs with the help of two good carpentry books and built a very nice set of stairs.

After the house was closed in, I wanted to work part time. I found no good-paying part time work such as my instrument repair work at the Music Emporium. I considered not hiring skilled workers for specific jobs as I had planned. If I paid a carpenter $15 per hour, I would have had to earn $25, before taxes, to have paid him for his hour. Since the jobs I could get paid only $6 per hour, I would have had to work four hours to pay the carpenter for one hour. It seemed to be a better deal to make the completion of my house my work.

This theory spread to all facets of the project. After the well, septic system, and foundation, I only hired Darrell Shedd to plumb the house. The labor to wire my house shifted from Bill Wilson to me as my ability and my homework assignments increased. I traded labor for all the additional help I needed. I taught Rick Rhodes guitar, and he helped me wallboard the ceilings. Jonathan Ginzberg, Richard Mansfield, and Dave Whitcomb always needed my carpentry skills. When they needed two men on a job, I went to their houses, and when I needed a second hand they came to my house.

Money for building the house and money for my living expenses came from the same checking account. I watched my house project money decrease with every dinner I ate out and every piece of clothing I bought. I was willing to spend the money for materials, but I was too frugal with my own life. Eventually, I had purchased almost all of the materials for the house. The materials sat in the upstairs rooms waiting to be used. I taught guitar lessons,

Our upstairs, master bedroom paneled in 1x6 tongue-and-groove spruce.

worked part time for Wilson as an electrician's helper, and earned some extra money from my land selling Christmas tree boughs for wreaths.

I gradually started a part-time business selling solar electric equipment. I obtained the appropriate dealerships for the major components. I had no money to advertise and no money to launch the business. I learned this new technology, and the customers found me by themselves. A customer would show up at my door sent by a neighbor or a friend. We would design a solar electric system; he or she would prepay, and I would deliver the equipment in a week or two. At the time, I could not see how this business would ever pay me a full-time income.

I continued juggling money from my part-time jobs and house project. Relief came when I enrolled in the Masters Program in Environmental Studies at Antioch New England Graduate School. I found that I could get a Guaranteed Student Loan for school and living expenses. I needed no credit rating and no job history. The money I borrowed and paid back in four years allowed me to smooth out my cash flow.

Two years after the beginning of the house project and the sale of my other houses, I sat down to do my income taxes. I had made a profit on my previous trailer and house. In order to defer all or some of the taxes on the profit, I had to account for the money I had spent on my replacement house on Bashan Hill Road. If the price to build the new house was greater than the price at which I had sold the previous properties, I could defer all the tax and pay no capital gains tax until I sold the new house. I carefully added all expenses. The total, which included land, septic system, well, labor, materials, and tools (including generator) was $41,000. There were changes and expansions over the next ten years that required additional funds. When I moved to Bashan Hill Road I had $28,000 in cash and the land that had cost $13,000. I had built my passive solar house on my nine acres of land within the limits of my $41,000 budget.

CHAPTER 14

The First Year: How the House Worked

URING THE WINTER of my first year at my new house, several friends approached Jonathan or Richard to ask, "How is he really doing out there? Is he okay?" Many people felt that I had fallen off the face of the earth when I moved that far back into the woods. People asked, "Why do you want to live way back there?" Many people thought I was freezing on Bashan Hill with the wind blowing through my partially finished house. Jonathan responded that my house was warmer than his or most anybody else's.

My house was not finished and decorated, but I did manage to get all the necessary steps finished by the time they were needed. The stove was installed in September when the nights began to get cold. The stove nicely heated the uninsulated shell in the fall. I finished insulating the house by Thanksgiving before the real winter weather set in. The stove easily heated the insulated house in even the coldest weather.

When it came time to insulate the walls, I found I had a problem with standard fiberglass insulation. Unfaced 4"x23" fiberglass insulation was not available. I could buy this size insulation only with standard vapor-barrier paper adhered to its surface. For the double-wall, staggered 2x4 construction of my outer walls, I needed unfaced four-inch insulation for the outermost wall and vapor-barrier-faced four-inch insulation for the interior wall. I could not use faced insulation in the outer wall because a second vapor barrier in the middle of the wall would have created condensation problems. My only solution was to buy faced insulation with the cheapest kraft-paper

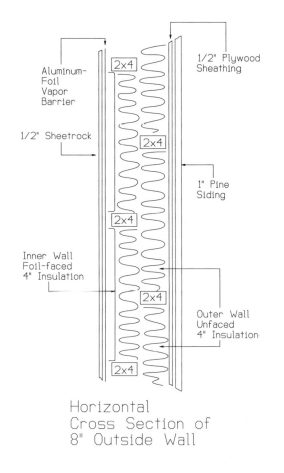

Aluminum-
Foil
Vapor
Barrier

1/2" Sheetrock

Inner Wall
Foil-faced
4" Insulation

1/2" Plywood
Sheathing

1" Pine
Siding

Outer Wall
Unfaced
4" Insulation

2x4

2x4

2x4

2x4

2x4

Horizontal
Cross Section of
8" Outside Wall

**Horizontal cross section
of double wall showing
the placement of
insulation and
aluminum-foil
vapor barriers.**

facing, remove the facing, and throw it away. I hated the additional handling
of the fiberglass.

I inserted the now unfaced insulation in the spaces between the 2x4s in the
outer walls. Next I installed vapor-barrier-faced insulation in all the spaces
between the studs of the innermost walls. The vapor barrier of this layer of
insulation was installed on the interior surface of the wall as in standard construction. For this faced insulation, I specially ordered aluminum-foil-faced
insulation. Aluminum-foil facing is a better vapor barrier than the standard
kraft facing. Additionally, a significant amount of the heat lost from a house
is actually radiated through the exterior walls. The shiny surface of alu-
minum-faced insulation reflects radiant heat back into the interior of the
house. Over the aluminum-foil-faced insulation, I placed a second, continuous vapor barrier of aluminum-foil-faced paper. This gave me two reflective
surfaces.

I insulated the ceilings between the first and second floors with six-inch
foil-faced insulation with an additional continuous aluminum-foil vapor bar-

rier. I wanted the heat from the large solar gain of the downstairs windows and the heat from the wood stove to remain downstairs and not rise too quickly to the bedrooms upstairs. The heat would have to circulate to the rear of the house and up the stairs before it could reach the bedrooms. The upstairs ceilings were insulated the same as the downstairs ceilings, with the addition of a second layer of six-inch insulation on top of the ceiling joists in the attic. The upstairs ceiling had a total of twelve inches of insulation. The greatest heat loss is always upward. I strategically placed the most insulation in this ceiling to counteract the phenomenon.

I also insulated most of the interior walls in the house. Fiberglass insulation is a good sound dampener. Furthermore, any room could be isolated heat-wise merely by shutting the door. At night I could shut my bedroom door and sleep in a cool room while the wood stove heat stayed downstairs where it belonged.

Before I installed drywall and paneling, the interior surfaces of my whole house, except for the floors, were all reflective aluminum-foil vapor barrier. I carefully duct-taped around each electrical box. Each box was backed by the four inches of insulation of the outer 2x4 wall of my double wall. This was far superior to the high heat losses of the electrical boxes of a conventionally insulated house.

I moved my living space or bedroom from room to room as I worked on the house. I started out living in the unfinished living room. While I finished the inside of the living room, I lived in the unfinished downstairs bedroom, my bed amongst the stacks of lumber. At one point, I remember friends of mine, Tim and Debbie, visiting while my queen-sized bed was in the unfinished kitchen next to the kitchen table. We had to use the bed as the couch.

The passive solar component of the house worked well from the beginning. As the weather got colder the winter sun was lower in the southern sky. The sun shone more directly into my south facing windows and farther into the house until it heated the whole surface of the slab. During even the coldest sunny winter day, the whole house was completely heated. At the end of the day when the sun dropped behind the westerly trees, the house would start to cool. I would start the Vermont Castings wood stove. I would heat with the wood stove through the evening until dawn. If the following day promised to be sunny, I would not fill the stove. If it was cloudy, I would heat the house with the wood stove in the absence of the sun.

I was also happy that the sun did not overheat the house. I had sized the thermal mass correctly. The mass stored part of the sun's heat. At night the heat that radiated from the thermal mass supplemented the heat from the wood stove. The house stayed a comfortable 70 to 75° F. during the sunny hours of a winter day. The upstairs had no thermal mass. I had reduced the south facing glazing in these rooms to about half of that for the downstairs to keep these rooms from overheating. The upstairs maintained a temperature similar to the downstairs.

At night the wood stove heated the living room to 80° F. This was the typical hot temperature of the wood stove room in a house heated with wood. The upstairs of the house did not overheat, thanks to the insulation between the floors and the circuitous path that the hot air had to follow to reach the upstairs. Hot air had to travel down a hallway to the rear of the house and up the stairs to reach the center hallway of the second floor.

My original house plan had called for window quilts on the inside of the south facing windows, which could be closed at night to reduce the nighttime heat loss. I have yet to install these window quilts or heavy drapes. I soon found that I needed only an average of two cords of wood to heat the house for the entire winter. The use of window quilts would save only a quarter of a cord of wood per winter. The purchase of the quilts and their installation has always kept drifting down the list of most needed improvements.

During the first fall, the Vermont Castings Vigilant wood stove adequately heated the partially insulated house. During the winter the stove easily heated the whole insulated house. However, when spring arrived, the stove produced too much heat. The house was tighter than it had been the previous fall when the insulation work was not yet completed. When I tried to run the stove at a slower rate, the wood burned too slowly, creating too much creosote from the poor combustion. Some of this smell would leak to the house. This stove was too large for my house in all but the winter months once the house was completely insulated. I later sold the Vigilant and replaced it with the smaller model, the Vermont Castings Resolute. This worked well in the spring, fall, or winter, but the smaller stove did not heat the kitchen sufficiently in subzero weather. A few years down the road I found a beautiful 1913 Glenwood cookstove at a bargain price. I installed this in the kitchen. I used this stove during cloudy subzero weather to heat the end of the house away from the main wood stove.

The house cracked and creaked during the first winter. The green lumber was drying in place. The beam in the living room developed long longitudinal cracks. When one of these splits let loose, a cracking noise would tear the silence. In the walls the green wood was shrinking, straining the nails. Rolo would jump out of a sound sleep barking, sure that the outside of the house was under attack from the hammer monsters.

The most pleasant surprise of the passive solar house was the feeling that I was almost outdoors and not completely indoors. People get claustrophobic in winter. Houses are dark, and cabin fever rises. This solar house felt like summer. There was a lot of natural light even on cloudy days. On sunny days I could sit in a rocking chair or lie on the floor and bask in the warm sun. I could literally close my eyes and pretend I was on the beach in Florida.

The first few times I left the house for the weekend, I drained the pipes to the basement. When I returned, I always checked the temperature. I soon had evidence that the house did not dip below 45° F. if I left it unattended. When the weather got bitterly cold, it was also exceptionally clear. The intense sun on these very cold and crystal-clear winter days provided maximum solar heat to balance the subzero nights. The house still did not drop below 45° F. in my absence, even in this extremely cold weather.

I had lived in other homes that were heated by wood. It was always a chore to go out for the day and evening. I would return to a house that had been unheated for many hours, since no one was at home to stoke the wood stove. When I came home late at night, I had to fuss with the stove for an hour before I could fill it for the night and go to sleep. In my new house I would return to a 60° F. house. If I chose to go straight to bed without starting the stove, the house would drop only to 55° F. by morning.

A conventionally insulated house not only drops in temperature quickly if left with no heat, it also takes a very long time for it to get back up to temperature. I once returned to my previous house to find that the backup furnace had failed while I was away for the weekend. The house was at 35° F. It took a whole day of heating with the wood stove to return the house to a normal temperature. My new house heated very quickly. If I was away for a week in the winter, I would return to a 45° F. house. I could shut the downstairs doors so the wood stove heated only the living room, the kitchen, and the bathroom. In half an hour I could heat the living room to 65° F., and in a few hours the downstairs was warm.

The cellar of the house had a wood stove that I used only when I was working on a project in the basement. At all other times there was no heat in the cellar other than the negligible heat loss from a hot-water heater. The cellar did not dip below 45° F. in the coldest part of the winter. The insulation on the perimeter stopped the heat loss through the walls. The floor was 50° F., the same temperature as the earth below it. The 50° F. floor and the warm house above actually heated the cellar to the 45° F.

The house was toasty warm and the natural light was great, but the house was very dark at night. As that first fall progressed, I had to stop work earlier and earlier, until daylight savings time and winter brought darkness at 4:15 p.m. I had refused to install the brighter gaslights to save my money for the future alternative electrical system. The only lighting I had was one kerosene lamp in the kitchen, one at the end of the couch, and one for my bedroom. The kerosene lights could not illuminate a whole room. If I wanted to read, I needed to hold the book just below the white globe that reflected the light downward.

Sometimes I would come home after dark and not be able to find the flashlight. I would then have to feel in the dark for the matches to light a kerosene lamp. There were other days when I would run out of matches. In this case I had to light a piece of paper from the gas stove to light the

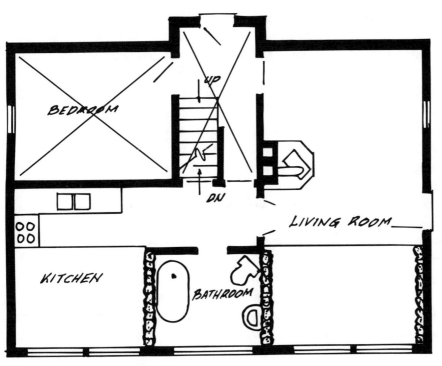

FIRST FLOOR PLAN

By closing two doors the majority of the passive solar heat and all of the wood heat could be confined to the living room, bathroom, and kitchen.

kerosene lamp. The burning kerosene also smelled. It seemed a contradiction to live in a healthful house and breathe kerosene smoke.

I RESEARCHED AND DREAMED of wind machines and alternative energy systems the whole first winter. In many ways this house project was a regression to a nineteenth-century lifestyle that was followed by a gradual return to a modern lifestyle. The installation of my small solar electric system the next winter was a significant leap forward. I probably felt as my rural ancestors felt when the Rural Electrification Act brought electricity to the farms of New England in the 1930s. I could now read in the evening and work on plans. I could find things. I could cook dinner with half the effort. I did not have to go to bed too early when my eyes were tired from the kerosene fumes and the poor light.

My first solar electric system powered one 12VDC light in the living room, one in the kitchen, and one in the bedroom. Each room had a lamp or fixture that provided general lighting for the room at the same time it provided reading light if I sat directly beneath it. I owned a small black-and-white TV left over from my father's belongings which could be powered by 12VDC. I wanted to play my records. I had read articles about others who were patching together 12VDC music systems. Radio Shack had manufactured a receiver that ran on 12VDC for use in recreational vehicles. An alternative energy magazine article detailed the modification of a specific Radio Shack turntable to run on 12VDC. Unfortunately, these items were now discontinued. I contacted all my local Radio Shack stores. No stores had these old models in stock, nor did they want to search for old stock.

My friend Jeffrey Hartman was in love and soon to be married. Jeffrey traveled most weekends from Boston to New York City to see Nicki. My brainstorm revolved around the premise that New York has everything you could ever want as long as you could locate it. I asked Jeffrey to see how many Radio Shacks were in Manhattan. Jeffrey called me from Nicki's apartment the next weekend to say that there were 27 and more in the boroughs. I gave Jeffrey the model numbers of the receiver and of the turntable. He called all of the stores and found several of each. These discontinued items were all discounted at various prices on the stores' close-out tables. We chose the best deals. The $200 receiver was reduced to $40, and the $140 turntable was reduced to $60. I had music.

I had no refrigeration for the first five years. I survived because I lived alone, and because I was a non–meat eater. Being alone helped because I only needed to juggle food for one person before it spoiled. Being a non–meat eater also helped. I ate more foods that needed little or no refrigeration. In the summer I had the large garden. All of my vegetables were fresh and needed no refrigeration. In the winter I had various cold boxes and the cellar to store food. In general, I bought food every time I was near a supermarket. I bought smaller amounts of food each time I shopped but shopped more often than I had before. One year I was a folksinger in a bar. I would hit the all-night supermarket after midnight because I would not be returning to the city for several days.

I almost had a propane refrigerator to use those first few years. I knew I needed to buy an old Servel propane refrigerator. Richard and the Cooks had them. Servels had been available in the past for less than $100. By the time I needed one, they had become scarce, as more of them reached the age of 30, 40, or 50 years. The price was up to $250. I would have paid the price except for the fact that these old Servels were reaching the end of their dependability scale. They lasted for decades, but by the early eighties they were an average of 40 years old. Many times someone would buy one that was healthy only to have it not work after taking it home. This was acceptable when Servels cost $50, but it was unacceptable as the selling prices increased to $250 or more. There were no new replacement parts and few old parts.

Bob Mason had a younger brother, Jeff. Through the grapevine, I learned that Jeff had a propane refrigerator for sale. I bought it sight unseen during the summer I was building the house. Jeff delivered it to my cellar hole, where it took four of us to unload it. It was the largest propane refrigerator I had ever seen. I did not know enough about gas refrigerators to know that they shifted from high to low flame as the thermostat indicated, to maintain the steady cold temperature inside of the box. Jeff explained that it needed repair because it only ran on one flame. I did not understand what he was trying to tell me.

The ugly pink refrigerator sat in the corner of the cellar for months. I did not have the time to learn about it. When I did look it over, I found it needed a lot of tender loving care. I was muddling along pretty well without refrigeration. I learned that Richard and Meg's old Servel burned 23 gallons of propane per month. This seemed like a lot for a little refrigeration.

One day, Rick and Sharyn Lafontaine arrived at my house. They introduced themselves and explained that Jeff Mason told them that he had sold his gas refrigerator to me. They were living in a small trailer in Cummington while they were building an earth-bermed home. They had no power and no refrigeration.

Rick and Sharyn drove home that day with the refrigerator for the same price of $100 I had paid Jeff Mason. Rick and Sharyn and I have been friends ever since. I installed their solar electric system, and they installed my front stone steps. They bestow a name on every piece of equipment they buy. They normally name a vehicle after the person who sold it to them. The old Ford tractor I traded to them got named Jeffrey. This refrigerator got named Big Pink. They eventually installed Big Pink in their trailer. The refrigerator worked well, but it only worked on the high flame. This meant that it always did the maximum cooling. The giant refrigerator used so much propane that they sold it three months later for the same $100. I am happy that Big Pink and I did not become better acquainted. I am happy she took her thirst for propane elsewhere.

The snow melted fairly early that first spring on Bashan Hill. I cleared the overgrown Christmas trees from the area due south of the house for my garden. The old garden was located in a corner of a field at the intersection of two stone walls. It had an unusually high number of boulders even for New England. I had decided to put the permanent garden where I cleared the trees. I left three-foot stumps, and I arranged for Ronnie Sampson to come over with his tractor and his PTO power winch to pluck them from the ground. In my cellar I had my new Rototiller to turn the soil into raised-bed gardens.

On Saturday, April 6, three days before Ronnie Sampson was to arrive with his tractor, a nor'easter arrived and deposited two feet of snow. This delayed my garden clearing. The following week Ronnie badly damaged his hand in a farm accident. My garden plans changed. I located the permanent garden where the temporary garden had been and spent the extra time necessary to remove the boulders.

CHAPTER 15

How an Alternative Energy System Works

A CONVENTIONAL HOME is attached to a power line that ultimately snakes its way back to a power plant. The power plant uses an engine or turbine, powered by propane, coal, diesel, hydro, or atomic power, to rotate a generator. The rotating generator produces AC, or alternating current, that circulates through the network of power lines to supply electricity to the users. The generator turns constantly, so the conventional home has electricity constantly available for its use. If the generator fails, there is no electricity until it resumes turning. For practicality, the power company has multiple generators such that if a generator stops, another is turning to ensure constant power.

A portable generator is the most common source of electricity to power a load at a site where the power company lines do not reach. As in the case of the power company, the electricity is supplied only when the generator is actually turning. Unfortunately, this type of supply is impractical for any remote home that needs electricity at all times of the day. The portable generator is not made to run 24 hours per day, every day. Furthermore, if this generator were to use half a gallon of gasoline per hour, it would run up a bill of $400 per month for the gasoline alone.

Any practical power source for a remote home cannot, both constantly and directly, supply power to the house and its loads. Since we would like to have lights when they are needed, not just when the wind is turning the wind turbine or the sun is shining on our solar electric modules, we need to store the energy when it is made, and

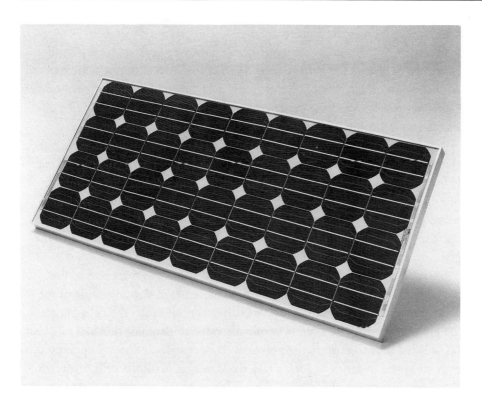

Solec International's model S-70 solar electric module. Photo provided by Solec International, Inc.

use the stored energy at our convenience. The solution is a battery or a group of batteries.

A battery is a device that changes electrical energy into chemical energy and stores it. It then converts the stored chemical energy back into electrical energy as it is needed. A battery stores electrical charge, which it discharges as DC, or direct-current electricity. DC electricity is much different from the AC electricity of a conventional home.

The center of any alternative energy system is the battery bank. The house draws all of its loads from the stored electricity in the battery bank. The longer a load is powered, the more electrical energy is used. Any amount of energy used is subtracted from the amount of energy stored in the battery bank. The bank supplies the appliances in the home with electricity from its finite stored energy. At some point the process of depleting the stored energy in the battery must be reversed, and the battery is recharged, so that it may continue to power more loads.

A simple alternative energy system operates in a manner similar to the 12V system of a car. When a car is turned off and the car stereo is working, the battery is supplying 12VDC electricity to the stereo. The longer the stereo runs, the more electricity is drawn from the battery. The battery starts at full and becomes less and less full the longer the stereo is on. Eventually the car is started. The motor turns the car alternator and the electricity that is produced recharges the battery. When the battery is full, the alternator ceases to charge the battery. If the car stereo is on while the car is running, the

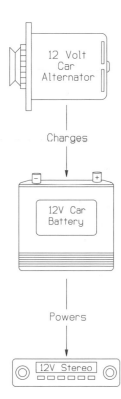

A simple alternative
energy system works in a
similar manner to the
12V system of a car.

alternator recharges the battery as fast as the car stereo depletes it.

In the 1930s, the most common alternative energy system used a wind generator to charge the battery bank. Today the most common source of electricity for an alternative energy home is an array of solar electric modules. Other sources are gasoline generators or hydroelectric generators. For our first example we will illustrate the workings of a solar electric system.

The electrical energy-producing device in a solar electric system is a solar electric module. A typical remote-home solar electric module is approximately four or five square feet. Conventional dimensions would be four feet by one and a quarter feet. The front surface that is exposed to the sun is tempered glass. This glass is mounted in an aluminum frame one to two inches thick to provide strength and integrity to the unit. Below the surface of the glass are approximately 36 four-inch square solar electric cells.

Each individual solar cell is extremely thin, about the thickness of the cover of this book. The cells are sandwiched between the underside of the glass and a plastic laminate. This plastic hermetically seals the cells between the glass and itself. Each cell is nominally rated to produce one half of a volt. (A volt is a unit to express the potential of electricity to flow.) The individual cells are wired in series in such a way that the sum of all the cells produces electricity in a range of 16 to 20 volts. This voltage is slightly higher than the nominal 12 volts of a 12V battery. The higher voltage is necessary for proper charging of a lead acid battery. A junction box is attached to the plastic laminate on the rear of the module. This box encloses two terminals, positive and negative, for wiring the solar electric module to a battery.

When the sun shines on a solar electric module, the rays pass through the glass and are absorbed by the dark color of the solar cells. Part of the sunlight provides energy to push electrons from the upper surface of the solar cell, across a barrier, to the rear layer of the cell. These electrons have a potential to return to the upper layer from where they originated. When a conductor, or wire, is connected from the lower layer to the upper layer, the electrons flow through the wire. Electrons moving through a conductor is the basic definition of electricity.

In a solar electric module, the individual solar electric cells are wired internally such that the module simulates the workings of the individual solar cell described above. When sunlight is absorbed by the cells in the module, electrons are driven across barriers and have a potential to return to the place from where they came. Additively, the solar electric cells create a nominal 12V solar electric module. Two terminals are located in a junction box on the rear of the solar electric module. One terminal is (−), or negative, and the other is (+), or positive. When a wire is connected between these two terminals, the electrons flow through the wire. If the (−) terminal of the 12V module is connected to the (−) terminal of a 12V battery, and the (+) terminal of the module is connected to the (+) battery terminal, the electrons flow from the module, through the wire to the battery, through the battery, and then back through the other wire to the module. The action of the electrons flowing through the battery charges the battery.

In theory, the simplest solar electric system consists of a solar electric module and an appliance such as a light bulb. If the module is placed in the sun, the module produces electricity and lights the bulb. When the sun does not shine, there is no electricity to light the bulb. We, of course, would most like to use a light bulb when the sun does not shine. A more useful solar electric system is composed of a solar electric module, a battery, and a light bulb. Now we can use the bulb when it is dark by depleting the stored electrical energy from the battery. The next day, the sun will shine, and the solar electric module will recharge the battery.

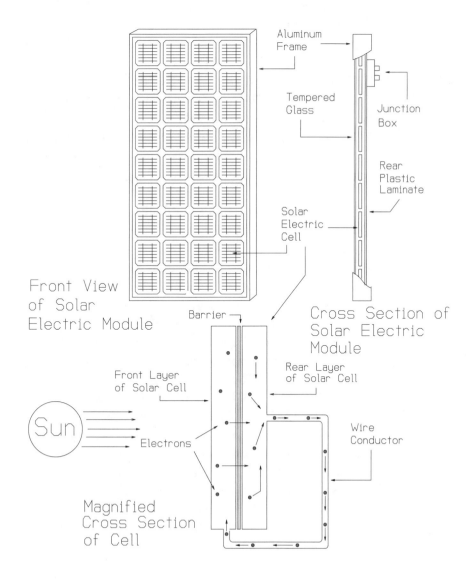

Front View
of Solar
Electric Module

Aluminum Frame

Tempered Glass

Junction Box

Rear Plastic Laminate

Solar Electric Cell

Cross Section of Solar Electric Module

Barrier

Front Layer of Solar Cell

Rear Layer of Solar Cell

Sun

Electrons

Wire Conductor

Magnified Cross Section of Cell

A solar electric module is composed of many solar electric cells sandwiched between a piece of tempered glass and a plastic laminate.

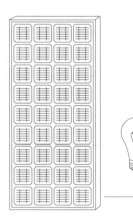

A single solar electric module directly powering a light bulb.

A battery holds only a finite amount of electrical energy. If too much electrical energy is put into a battery, it will overcharge. This can damage the battery or reduce its life. For this reason, it is prudent to add a device called a "charge controller" between the solar electric module and the battery. This device effectively disconnects the solar electric module from the battery once the battery has reached full charge.

The smallest solar electric system may have only one solar electric module, a charge controller, and one battery. This system produces only enough electricity in the winter in New England to power one or two efficient 12V lights in the evening. For larger systems with greater needs, one uses more solar electric modules for more energy production and more batteries for more electrical energy storage.

Solar electric systems for remote homes normally operate at 12VDC or 24VDC. Solar electric modules are manufactured in a 12V configuration. It is easy to wire pairs of 12V modules into 24V sets. These pairs can be paralleled to make a larger array of modules at 24V. The one battery of our small system operates at 12 volts. Many batteries may be wired together to create a battery bank at either 12 volts or a 24 volts. The batteries, or battery banks, produce only DC electricity such as the 12VDC electricity in a car. There is a selection of 12VDC or 24VDC appliances available from the recreational vehicle industry and the marine industry. Higher voltage systems are possible but uncommon because few appliances that run on higher voltage DC are available.

Smaller solar electric systems produce, store, and utilize low voltage DC electricity only. All of the appliances in the cabin or home are specifically selected to operate on low voltage DC. As the size of the solar electric system increases, and the selection of appliances expands, it usually becomes necessary at some point to use some standard household appliances. To power conventional alternating current (120VAC) appliances, we need a new component in the solar electric system: an inverter.

An inverter is an electronic black box that changes DC, or direct-current electricity, into AC, or alternating-current electricity. Because of the popularity of 12V and 24V systems in RVs, boats, and alternative energy homes, most inverters change 12VDC or 24VDC electricity into 120VAC electricity. These inverters are 90% efficient in their energy transformation. Our solar electric system now has multiple solar electric modules, a charge controller, a battery bank as a source of low voltage DC electricity, and an inverter to produce standard 120 VAC electricity.

A customer of the power company has little understanding of the amount of energy he or she is consuming. The juice flows in through the meter and the monthly checks go out in the mail. Most people cannot read a meter, nor are they aware of the electrical demand of any given appliance. I sometimes think the system is set up this way on purpose. What if it were more like gaso-

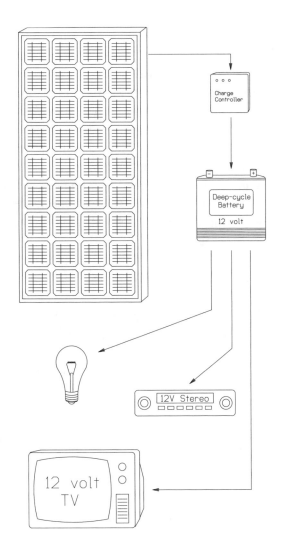

A one-module solar electric system.

line and your car, and you could check the miles per gallon? The supply of electricity for the homeowner's use is relatively limitless. A homeowner can use $50 worth of electricity one month and then use $100 of electricity another month when the mother-in-law is using the electric heat in the guest room.

In an alternative energy home, there is a finite average production of electrical energy. If there are only ten solar electric modules, then they will average only about two kilowatt-hours of electrical energy per day in the summer. Devices such as trackers and mounting angle adjustments can increase the average production only. There is no way to increase the production for a specific period of time, such as the visit from the mother-in-law. You cannot just pay $50 more for that month. If you want more electricity, you need more modules. When you buy more modules, you do not buy extra electricity for the month; you invest in a bigger power plant to give yourself more electricity every month for the next twenty years.

Normally the size of an alternative energy system is determined by the loads that must be supplied and by the amount of money that the buyer can afford to invest. In New England, where we live, the average solar electric energy production in the summer is twice what it is in the winter. Furthermore, the winter energy production can be sporadic. In the worst winters we receive virtually no sunny days for three consecutive weeks during November or December. In normal years, the sun appears for a sunny day once a week during that same spell. We learn to adapt to the irregularities of electrical energy production. One of our solutions is to store "negative electrical energy use" in the form of dirty laundry. During a bad spell of cloudy weather, we save the non-essential dirty laundry for a sunny day to smooth out our energy consumption.

The finite amount of energy makes one think about the energy budget. A large, electrically demanding construction project should be scheduled for the best time of the year for the electrical energy budget. The yearly washing of drapes and blankets should be done other than in the cloudy spells. A work project that demands all day computer use may not be scheduled for the worst weeks.

Another consideration of a solar electric system, besides the amount of energy that the system produces, is the amount of energy the system stores. When the sun shines every day without fail, as it does in the southwest of the

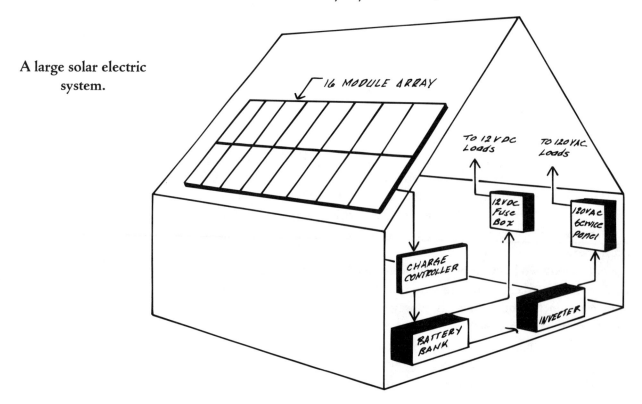

A large solar electric system.

A Trace 2,500 watt inverter. Photo provided by Trace Engineering Company, Inc.

United States, the battery bank can be small. When sunshine is sporadic, as it is in New England winters, the battery bank should be large. The large battery bank allows the household to power many days of loads while waiting for the next sunny spell to recharge the large deficit in the battery bank. The large storage of a large battery bank does not increase the amount of electrical energy produced by the solar electric modules; it only smooths out the monthly budget of energy.

Our detailed example of an alternative energy system has been for a solar electric system. Other alternative energy systems are actually extremely similar in structure.

A wind powered system has almost all of the same components of the solar electric system. The one major difference is the substitution of a wind generator for the solar electric modules. When the wind blows, the propellers rotate a shaft, which in turn rotates a generator. The rotating generator supplies low voltage DC electricity to a charge controller, which governs the charging of the batteries. A wind machine is dependent on good exposure to the wind at a good wind site. This normally means a high tower. The solar electric system needs good exposure to sun at a good solar site.

A home sized hydroelectric system captures the energy of moving water to rotate a turbine that, in turn, rotates a generator that supplies low voltage DC electricity to a charge controller. The rest of the system is the same as a solar electric system. The novice thinks of a hydroelectric system as a waterwheel in a river. This type of water flow is hard to harvest for a home alternative energy system. A home hydroelectric system is normally composed of a pipe two to four inches in diameter that channels water from a point 50 to 200 feet higher than the turbine, down to the turbine. Preferably the turbine is close to the battery bank and the uphill water source is the distant part of the system. The strength of a home hydroelectric system is its consistent production of electricity. The home hydroelectric system produces a small amount of electricity per hour, but it produces this electricity all day every day.

An alternative energy system may also be powered by a gasoline or diesel generator. The generator produces 120 VAC current to power a battery charger to charge the battery bank. The rest of the system is again similar to the solar electric system. The generator may be turned on manually, or it may be activated automatically whenever the batteries reach a specified low state of charge. Most home-sized inverters come with a battery charger option that greatly simplifies the

An alternative energy system may also be powered by a wind machine, a hydroelectric generator, or a diesel or gasoline generator.

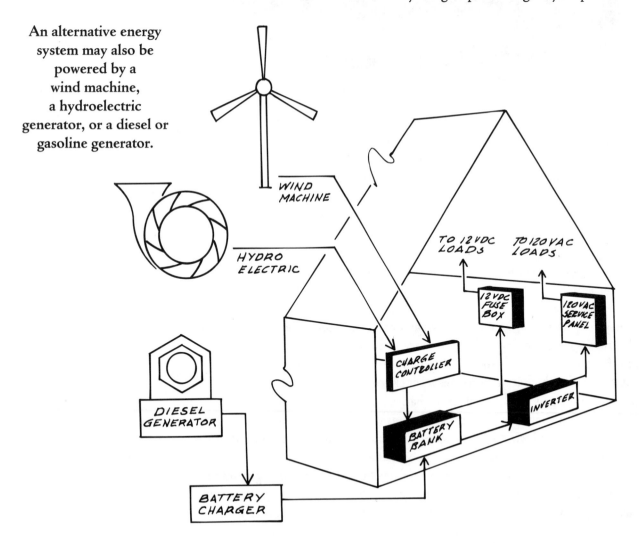

system. When the generator is started, the inverter ceases to be an inverter and effectively runs backwards as a large and sophisticated battery charger. Specific large AC loads may be powered by the generator only.

CHAPTER 16

Chasing Down Windmills

I HAD ALWAYS THOUGHT I would purchase a wind machine as my source of alternative energy. I like to listen to the wind blow in these Berkshire Hills. It would be pleasant to listen to the wind on a winter's night knowing that it was powering my home at the same time.

The alternative energy books that I read previous to my move to Bashan Hill devoted the most space to the subject of wind power. At the time when these books were written, wind was the best source for alternative energy power if the site of the home was at a good wind site. Many of the books discussed the Savonius vertical axis wind machine. The simple turbine was easy to build. It could be coupled with an inexpensive car alternator. The Savonius turbine turned slowly but forcefully. This characteristic made it more suitable for water pumping and less suitable for electrical production. I spent many hours dreamily designing wind machines in a graph paper notebook. Like all the engineers who were much more advanced than I, I never came up with a better design than the standard propeller wind machine.

Locally, there was a strong resurgence of interest in wind power in the early 1980s. A professor at the nearby University of Massachusetts had started a business selling and installing wind machines. Western Massachusetts had a large population of counterculture and old Yankee do-it-yourselfers. These people took the oil crisis of the early 1970s very seriously. They abandoned the oil companies for wood heat. There were organizations for energy efficiency. At get-togethers favorite topics of conversation were not real estate, cars, and possessions, but wood stoves, energy efficiency, and insulation. This was a time when most large bookstores had significant sections dedicated to energy efficiency and efficient home design.

Nationally, there was an energy tax credit. This was most known for its subsidy of solar hot water and insulation retrofit. The same law also covered electrical energy production from renewable energy sources such as wind and solar. The tax credit worked this way: if you purchased $1,000 worth of equipment, you got 40% of that amount, $400, off the taxes you owed the federal government, to be taken that year or over the next three years in increments. The 40% was good for purchases of up to $10,000. Selected states also had tax credits for the state taxes. Massachusetts had a tax law that effectively gave 20 percent with a $1,000 maximum credit per principal residence.

The majority of the wind machines that were installed at this time were installed at homes that were connected to the power line. There was also a law in Massachusetts that effectively allowed a homeowner to produce power and turn his meter backward when surplus power went backward to the power company. A homeowner could purchase a $10,000 wind machine. He or she could then take $5,000 in state and federal tax credits. The wind machine now only cost $5,000. Unfortunately, the wind machines were still not cost effective for the homeowner. The machines did not produce in the field as they had been expected to when considered on the drawing board. Most of them were taken down over the next few years. These machines could never be purchased for alternative energy homes because they were manufactured to produce 120 volts and interface with a synchronous inverter connected to the utility company.

When I first stopped to meet Bob and Karin Cook, Bob and I instantly got into a discussion of wind machines. Bob had been following the product information for years. He felt that the area was a good wind site. Wind power was really the only consideration for us at Bashan Hill at the time. The only other theoretical consideration was hydropower. We were up near the top of the hills. We were located at the source of streams rather than at a point where the water could flow by us from above and power a hydroelectric turbine.

My property on Bashan Hill was theoretically a good wind site. Bashan Hill was the highest hill in Worthington. My property was half a mile to the west of Bashan Hill. The next area higher was another half a mile to the west. My property was only 300 feet lower in altitude than either Bashan Hill or the ridge to the west. To the south and southeast the land sloped down and away. To the north the land maintained the same altitude as my land.

My property was mainly covered with 25 foot tall overgrown Christmas trees. Only to the north on Bashan Hill Road did the tree line get higher. These trees were 60 or 70 feet tall. In general, a wind machine must be

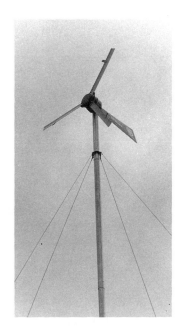

A Bergey 850 watt wind machine.

installed at least 30 feet above all obstacles closer than 200 or 300 feet. In my case the high trees lining the road were 250 feet away. In all other directions the tree line was no more than 30 feet. This meant that theoretically, a wind machine could be mounted on a 60 foot tower.

Many people believe that a wind machine could be powered at their site because they have windy days. In actuality, the average wind that blows hard by a house, and over and around buildings and trees, will not power a wind machine. The wind is actually much gustier than it appears. A wind machine in this type of wind will turn faster and slower with gusts, and it will constantly change directions as it tries to stay in the wind. A good electrical output from a wind machine is dependent on a strong, steady wind from a constant direction. Imagine blowing a long, steady lungful of air at a pinwheel. It will gradually gain speed and maintain a steady RPM. If you instead puff several blows at the pinwheel, it will not reach and maintain the same fast revolutions.

One of the best places to install a wind machine is on a high tower on the top of a mountain. The land is far below. The natural turbulence of the wind passing over obstacles is confined to an area close to the ground. The wind that is passing by the turbine up at the top of the high tower is steady. Another good site for a wind machine is on the shore of an island or lake, where the prevailing winds come off the body of water. The smooth surface of the water causes little turbulence in the wind. In this case a lower tower is possible.

All things considered, my Bashan Hill property was, and still is, a potentially good wind site. Through the first winter at Bashan Hill, I sent for and studied information on wind power. This was the predominant subject of conversation for me and Bob whenever we got together. Bob was most strongly considering a one or two kilowatt wind machine. I was looking at smaller machines that were less expensive and, correspondingly, could be installed on a lighter weight, less expensive tower.

One day in the fall of my first year at Bashan Hill, I found an advertisement in the local newspaper selling a wind machine. I drove to Easthampton, a town bordering Northampton, where I shopped for food. The man was my age. He was living in an apartment, saving money to move to his piece of land up in the Adirondacks of New York. He had bought the wind machine for his remote land. Now he needed money. He had decided to sell the wind machine because he first needed to buy building materials.

The wind machine was a 500 watt, 12V Sencenbaugh wind machine built by Jim Sencenbaugh from California. I was familiar with Sencenbaugh's wind machines. The machine was brand new and still in the original boxes. The asking price was $1,000. Originally the man had paid $1,500. The owner wanted to sell the machine badly. He was not very interested in dickering on price, but he was interested in a partial trade. He needed building

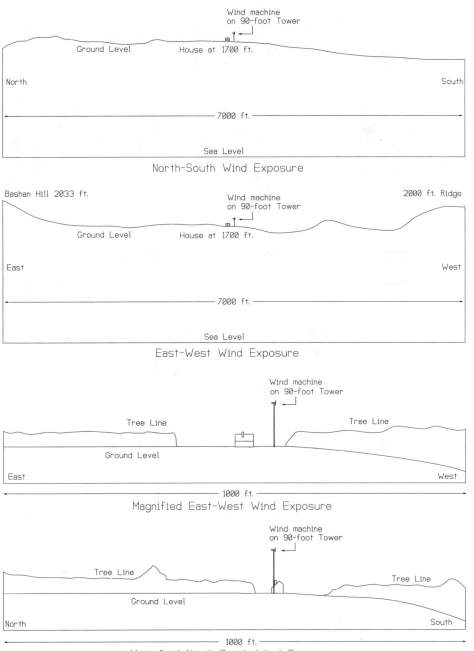

Wind machine
on 90-foot Tower

Ground Level   House at 1700 ft.

North   South

7000 ft.

Sea Level

North-South Wind Exposure

Bashan Hill 2033 ft.   2000 ft. Ridge

Wind machine
on 90-foot Tower

Ground Level   House at 1700 ft.

East   West

7000 ft.

Sea Level

East-West Wind Exposure

Wind machine
on 90-foot Tower

Tree Line   Tree Line

Ground Level

East   West

1000 ft.

Magnified East-West Wind Exposure

Wind machine
on 90-foot Tower

Tree Line   Tree Line

Ground Level

North   South

1000 ft.

Magnified North-South Wind Exposure

**The wind expo-
sure at my house
site for a poten-
tial wind machine
installed on a 90
foot tower.**

My neighbors Bob and
Karin Cook.

materials. I had no extra lumber, but I did have Big Pink, the old propane
refrigerator. We talked about $800 and the refrigerator. I was interested. I
took the man's telephone number and returned home to think about it.

I telephoned Jim Sencenbaugh in California to get some more specifications. I needed to know recommended wire sizes for the long run of wire
from the top of the tower to the batteries in the basement. Voltage drop in a
12V long wire run is always a potential problem. Jim told me the machine
ran at a higher voltage to compensate for line losses. Six-gauge wires were
appropriate. Jim also gave me the strength rating for a tower for this wind
machine. When I priced new towers I found they were more expensive than
the machine itself. There was a possibility that I could buy a used tower from
the wind machine store in Hadley, but the prospective tower had not actually
been traded in yet.

A few days later I saw Clarence Witter at his sawmill. I told him about the
wind machine. Clarence was always excited about new ideas. He quickly told

me he had a woodlot he was cutting which had very tall and straight red pines. He offered to cut poles and deliver them to me for a pole tower and braces. He would even use his cherry picker to get the tower pieces in an upright position during installation. Everything was falling into place. I called the man up to negotiate a final price for the wind machine. I left a message that I was interested in buying the wind machine but he did not return my call. I naturally assumed he had sold the machine or had changed his mind.

A month later I received a call back from the man with the wind machine. Apparently he had lost my number. I would have bought the wind machine originally, but critical time had passed. I told him I would reconsider. I never called him back. Somehow the ducks were no longer all in line. Winter was beginning. Any installation would have to be put off until spring. I decided to wait.

I CONTINUED TO THINK about wind machines. They were still really the only consideration, but I never felt secure with them. I was leery of entering into an expensive adventure that could turn into a downward spiral. I still do not have a wind machine, though I know it would provide a balance to the low winter output of our solar electric system.

There were distinct problems with wind machines in 1982; some have been solved, but many of these problems still exist today. Some have been improved. The biggest problem was the uncertainty of the actual power output of a wind machine. There were several wind machines in the area. They were not doing well. People were uninstalling them. When I drove past these machines, they were normally standing still. The books and wind maps indicated that there was enough wind to supply an ample amount of electricity to my home. This data did not seem to match with what I observed. Was my wind site that much better than these other sites?

AS I DELVED DEEPER into the wind power magazines and books, I concluded that the only way to truly evaluate a wind site was to set up a wind sensing device to record the output at the site for a year. Unfortunately, this apparatus was expensive. It was just as economically feasible to set up a small wind machine and find out firsthand. Bob and I started to keep a casual mental record of the wind we observed here. As the winter ended and summer arrived, I could see that the wind simply did not blow often enough to power my home in the summer. I would need to install a generator and battery charger to survive in the summer. The obvious question was: Why not make electricity with a generator and battery charger all year round, once I had purchased the generator, battery charger, and the battery bank, and forget the expensive wind machine? I continued my consideration of wind machines largely because of the tax credits.

Even if this was a good wind site, there were still many other problems.

The wind machine would need a very large battery bank to get through the long periods when there was no wind. If I was to install a wind system, I would eventually want to power my deep-well pump. This was a necessary everyday load. I would need extra water during the summer, when the wind would not blow, to water my garden. Batteries need to be returned to full charge relatively often. A long summer of intermittent wind would mean a constant rationing of electrical use. I would be praying for wind to recharge my expensive battery bank to full before it was irreversibly damaged.

The selection and the installation of a wind machine seemed to be a very rigid proposal. I would have had to choose the voltage, the size of the machine, and the tower height at the start of the project. I was used to projects that allowed me more control of the variables. I normally had to make changes in a project several times during its beginning stages. In this case I would have to make the correct choices in the beginning. If I wanted to change from a 120V wind machine to a 12V or a 24V wind machine I would have to sell the old machine and buy a whole new machine. If I wanted more power, I would have to sell the old wind machine and purchase a larger one. The new larger machine, of course, would need a stronger tower.

I would have needed at least a 60 foot tower and possibly a 90 foot tower. These towers could not be easily lowered once they were in place. To install and later to maintain the wind machine, I would have to climb the tower. If the machine needed any real service, I would have to climb the tower, dismantle the machine, lower it to the ground, then later climb the tower, hoist up the wind machine, and reinstall it. I was not afraid of heights. I had worked on ladders and high roofs. I was willing to climb one of these towers. I just could not be sure that I would want to climb one these towers when I was much older.

We had several ice storms the first year on Bashan Hill. The storms that brought a slight icing to the rest of the town were more severe at our higher altitude. I had read that ice buildup on the blades of a wind machine could unbalance the blades. If the winds were high, the imbalanced blades could vibrate a wind machine to a point of failure. I could have climbed a high tower, but I did not want to worry about climbing a tower during an ice storm to protect my large investment. Wind machines also sometimes lose their overspeed protection and spin into oblivion. I had read of a six-foot wooden blade crashing through a roof.

A wind machine is a potentially high maintenance machine, as is any machine with rapidly moving parts. Bearings wear out. Grease seals fail. Parts go out of balance setting up destructive vibrations. At the time, the average machine required yearly lubrication. If the machine failed, it would have to be returned to the factory for repair. This would be a long process. Many experts felt that the old machines from the 1930s were the most reliable. The Windcharger 200 watt model was the only old wind machine still manufactured. This was too small to supply adequate power to a whole household.

The most attractive reason to install a wind machine was the large potential output estimated in wind power books. I have never felt sure that this potential would really come to fruition. There was plenty of local evidence that it would not. I could never adequately answer the many negative questions I had concerning a wind machine installation. I was glad that fate had prevented me from purchasing the Sencenbaugh 500.

CHAPTER 17

My First Solar Electric System

T HE ALTERNATIVE ENERGY books that I had read previous to my move to Bashan Hill either did not mention solar electricity, or they mentioned solar electricity as a possible energy source of the distant future. These books considered solar electricity to be a specialized exotic source of energy for the satellites in outer space. The books I was reading were published in 1979 or before. As with most resource books, the information was actually a year or more older than the printing date.

During the first winter I received some literature from Richard Mansfield about solar electricity. He said, "Maybe you will be interested in this." What he really meant was, "Do you think this is any good?" I wish I could tell you that I jumped on the bandwagon for solar electricity at that point. I told Richard that it was worthless. I threw the material away and went back to wind machines. A few months later I made a call to a new alternative energy store in the Greenfield area. I wanted to know what they recommended for wind machines. They were not recommending wind for independent homes. They were recommending a new energy source that was gaining popularity in California — solar electricity. I politely argued with them and just as politely discounted what they had to say.

I was making two mistakes when I threw away the information I had received from Richard. The first mistake was believing the estimates for the amount of energy that could be produced by a small wind machine. The second mistake was assuming that I needed a lot of electrical energy at my home in the first place. I had recently moved from a similarly sized home. I had had a conservative approach to the use of electricity. That home averaged a monthly consumption of only 250 kilowatt-hours. A wind machine promised me an average of 125 kilowatt-hours. I felt that I could cut my use of electricity in half and meet the budget. The solar electric product information I had received from Richard seemed to indicate that an investment in solar electric modules would produce only a fraction of the yearly output of a similarly priced wind machine. Buying solar electricity looked more like buying a science project toy than buying an alternative energy system.

Late in the summer of 1982, Bob Cook loaned me some information on solar electricity. Part of the information was the identical material I had received from Richard. It explained how to become part of the cooperative purchase of solar electric modules organized by a man in Arkansas. The other part of the information was a government study of the recent use of solar electricity in independent and remote homes. Bob made three clear and simple points while he was handing the material to me. 1) The solar energy of New England is not much less than that of California except for a few months of winter. 2) The sun shines a lot more regularly than the wind blows. 3) I should read the testimonials in the government publication describing the loads these people were powering with small solar electric systems.

Bob's first point was well known to me. I had all of the solar statistics memorized from the project of my passive solar house. The second point defined the reason why I had been ambivalent about purchasing a wind machine. I never felt certain that the wind would blow hard enough and often enough for a wind machine to produce anywhere near its estimated output. Today, after many years of experience, I can easily define what the problems would have been for a wind system. Even if the estimates had been correct, I would not have been able to harness anywhere near the required output from the wind machine.

The wind machine would have produced energy intermittently. The long no-wind periods would have been a serious problem. What is not as obvious is that most of the yearly estimated output of the wind machine would not have been usable because the batteries would have been fully charged. During a windstorm, the battery bank would have been fully charged during the first stages of the storm. The second part of the storm would have turned the wind machine, but my system would have had no place to store the power it produced. Moderate winds contribute only a small amount to the total yearly output of a wind machine. High winds produce the majority. Much of this high wind energy production and, correspondingly, a large part of the yearly production, would have been wasted because it would have occurred at a time when the battery bank was fully charged.

The testimonials of the people who were using solar electricity were the most important argument for solar electricity. Many of the people powered their homes with one to four 33 watt solar electric modules. Each module produced about 200 watt-hours of electricity per day in sunny California. The solar electric modules were virtually maintenance free. They had no moving parts and an estimated life of twenty years. Many of the people were considering adding more modules when they could afford it. The expansion of a system required no more than an additional mounting structure and some additional batteries. An individual's energy system could be easily expanded in response to changes in use and finances.

Each example detailed not only the solar electric equipment but the appliances that were powered. I was encouraged that a very small system could greatly impact one's quality of life. A typical home with a two-module system was powering a few 12V lights, a 12V stereo, a 12V TV, and a 12V fan. These were my most basic needs. Other people pumped water with a 12V pump. Generators provided the 120VAC power for tools and washing machines.

These people were similar to me, the Cooks, and Richard and Meg. They had advanced from the most primitive level of kerosene or gaslights. We would be happy, as they had been, with incremental improvements. From the description of the appliances that they used, I could see that efficiency and cleverness were as important as the solar power. Instead of buying a particular wind machine of a specific size on the correct tower, which could meet the average demand of a whole house, these people were buying a small amount of production and cleverly using all of the energy produced. I had planned my house with many ideas of conservation. I had eliminated many large electrical loads. These people were looking at the power use of each individual load, no matter how small. They were specifically choosing the 15 watt reading lamp over the 50 watt general-lighting ceiling fixture.

I sat down with the solar electric product information to design a small system for my home. I did not attempt to make a solar electric system compete in output with a wind system. I estimated the small amount of power I would get from a small system. Next I tried to estimate the combination of loads that I could power from this small system. I felt a solar electric system with two 33 watt modules might produce enough electricity to power a light in the kitchen and a light in the living room. I could also power my 12V black-and-white TV. This only used 15 watts. The long winter periods of no sun would be the real test. I felt that I would be safe in the worst times because I could cut my electrical use to only one light. I was one person. I could light only the space I was in at any given time. I could watch the TV with no background light.

By fall of 1982, Bob and I were committed to trying solar electricity. Bob got his order organized before I did. One day he announced that he was sending his check off in the mail. My first reaction was to rush to the bank and send my check with his. I confess that I let Bob be the pioneering guinea

Joel Davidson in the early 1980s. Photo provided by Joel Davidson.

pig. I waited for his order to arrive. After all, Bob was sending a $750 bank check off to some backwoods guy in Arkansas for a cooperative buy. It was always in the back of my mind that the money would disappear into a black hole. Mostly I think I just wanted to see and feel a module first.

Bob flagged me down as I drove past his house one day to announce the arrival of the solar electric modules. I could not believe how good they looked and how well made they were. I made my decision on the spot. I would also order two modules.

JOEL DAVIDSON was the person behind the cooperative sale of solar electric modules. A year later I got to know Joel by telephone, and in later years we have become good friends. He has been a business visitor and at other times a social visitor. In 1982, Joel was living in the hills of Arkansas in a cabin with no power. Joel was quite a character. He came out of the streets of

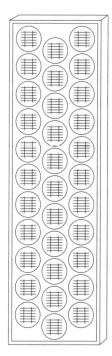

**One of my first two solar
electric modules.**

Philadelphia to later end up as a sixties back-to-the-lander. He fell in love with solar electricity and its potential to provide independent living for people like him and to save the world from the pollution of the other methods of electrical production.

In the late 1970s, solar electric modules were not available for the individual to purchase. These modules were produced and used for specific commercial and military applications and sold only in large lots. Joel was, and still is, a great salesman and a great pitchman. This is his talent. He can speak sincerely, passionately, and knowledgeably about solar electrically saving the world.

Joel, the organizer, had set out to purchase solar electric modules. He convinced Wm. Lamb Corporation to sell him modules. Joel had negotiated a minimum order well below the normal lot size. Joel then advertised in a few alternative energy magazines for people interested in a cooperative purchase of solar electric modules. The first information I had received from Richard (and discarded) was actually from Joel. Joel collected money orders and bank checks from small purchasers until he reached the threshold of a minimum order for the distributor. He then sent in all the money, and Wm. Lamb Corporation shipped the modules. Most cooperative purchasing occurs because people want a better price. In this case, the cooperative purchase supplied a product that was not otherwise available. It is amazing that all of this actually worked, but it did. Joel wrote and distributed a small, home-grown newsletter for solar electric users, the *PV Network News*. He had almost singlehandedly created a solar electric movement from the outback of rural Arkansas.

In the fall of 1982, I did not know who this Joel Davidson was, or if he even really existed. I felt better when I saw that Bob Cook's modules had actually arrived. I prepared my order and procured my $750 bank check. I could not bear to just send this money off in the mail. This was a time before the popularity of mail order. I wanted the assurance of hearing his voice. I called, but Joel was not home. Instead, I reached Joel's twelve-year-old daughter. She was helpful, kind, and reassuring. She carefully took a message. I did not wait for a return call. I was satisfied. I mailed my check. My two 33 watt modules arrived two weeks later.

One of Bob Cook's friends had worked in the engineering of an electric vehicle. This friend recommended an Exide deep-cycle 6V golf-cart battery for our solar electric systems. Bob got his batteries through his friend. I found my batteries at the nearest Exide distributor, 120 miles away in Boston. This was not too much of a hardship because I always wanted an excuse to visit friends there. The distributor would only accept cash. I also had to become a dealer to get a decent price.

The average two-module solar electric system would have employed two of these 6V deep-cycle batteries. Bob and I reasoned that a battery bank of four batteries would greatly help us average out the uneven production of electricity in the low-sun months of winter. The match of two 33 watt modules to the four batteries resulted in one half of the charge rate per battery. At the lower level of amps going into each battery there was no danger of damaging the batteries from overcharge. This meant I would not need to purchase a charge controller. At the time, a simple and small charge controller cost $100.

Every facet of this little solar electric project was determined by cost. Dollars were scarce, and this was a new and unproven venture. If all failed, I wanted to have spent the minimum. Furthermore, all independent homeowners were approaching solar electricity this way. Most of the advice in the *PV Network News* addressed the ideas of how to install a solar electric system for less. Most of this attitude came from the fact that the solar electric modules were so expensive that no one wanted to spend money on anything else. People shared experience on how to salvage electrical parts, light sockets, and bulbs from junk cars.

The easiest part of the solar electric system project was purchasing the batteries and the modules. The hardest part was obtaining the small parts needed to put the system together. I had no source for battery cables. Luckily, I had my friend Bill Wilson, the electrician. He saved me leftover short pieces of heavy copper cable. From the electrical distributor he brought me heavy-duty bolt-on lugs. By the fall of 1982, I had learned enough from Bill Wilson while wiring my home to work as his assistant on his jobs. Every day we used the wiring, the electrical boxes, the receptacles, and the switches that were required by the electrical code for home installation. In the evening I looked at the RV and auto industry electrical equipment that was recommended for small 12V alternative energy systems. The equipment for the home was far superior. Because the home equipment was in mass distribution, it was also much cheaper. With the help of Bill, I bypassed most of the RV and auto electrical equipment.

My small solar electric system cost about $1,100. The modules cost $750, the batteries cost $270, and the small parts and frame cost about $80. Massachusetts had a law that declared equipment purchased for use in certain renewable energy systems, such as wind or solar, sales-tax free, saving me $55. The federal government had a 40 percent tax credit for the equipment used in a solar electric system. Massachusetts had a state tax credit that amounted to 21 percent of the price of the system. After I had taken all these tax credits over the next few years, the solar electric system really only cost me $429.

Installing my first solar electric system was about as easy as inventing the wheel. I had no real instructions. I did have the theory of a 12V alternative energy system from the books that I had read. I started with a pile that consisted of the two modules, the four batteries, two 12V, 25 watt light bulbs, and a 12V black-and-white TV. If I was doing the project today I would fill

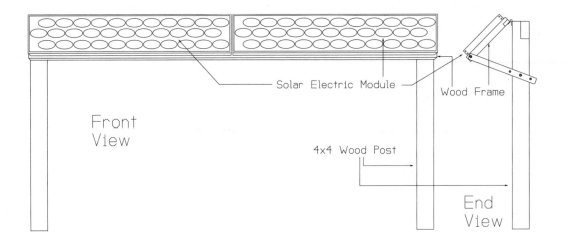

Front View

Solar Electric Module

Wood Frame

4x4 Wood Post

End View

My first solar electric modules were mounted on two 4x4 posts in the ground.

out an order for a mounting frame, battery cables, wire, charge controller, and all the other appropriate pieces, then send a few hundred dollars to a mail order company. The goods would arrive in a few days. Even as a novice, I would have the system installed in a relatively short time. In 1982 I had to think about every piece and hunt for every part.

I started with the mounting frame for the modules. I had many choices for mounting the modules because I had an exceptionally good solar site. I chose not to mount the modules on the roof or the upper south side of the house. I was sure the installation would only be temporary. If the solar modules functioned poorly, they would come down. If they operated well, I would add additional modules and a new mounting structure. I did not want to patch holes in the roof or on the side wall of the house. I therefore chose to mount the modules on a ground mount south of the house. The modules would be easily accessible for any testing or repair. I could clear snow from them in winter and adjust their angle seasonally.

The modules that I had purchased measured one-and-one-half-inches thick by one foot by four feet. These modules were, if anything, overbuilt. In future years the aluminum frames of modules would become lighter. These modules were strong in their own structure. A mounting structure's only function was to provide a sturdy frame on which to attach the modules, which were rated to withstand 120 MPH winds and one-inch hailstones. My mounting structure was made from wood. I was a carpenter, not a welder. I made a simple frame of 2x3s. The frame was one foot wide by eight feet long. The two modules butted end to end.

The most difficult part of mounting the modules was finding the correct nuts and bolts. The module frames were made of aluminum. Standard nuts and bolts are steel. Wherever steel and aluminum touch, a chemical reaction causes the aluminum to oxidize. I did not want my expensive modules to corrode and deteriorate before I used them for their full twenty-year life. I needed aluminum bolts and nuts. The only place I knew to buy aluminum nuts and bolts was from the assorted-aluminum-nuts-and-bolts box at the hardware store. I traveled to several stores before I could find all of the eight nuts and bolts that I needed for the project.

This frame was mounted on two 4x4 pressure treated posts, twenty feet south of the house. The poles were oriented in a line from east to west, eight feet apart such that the frame was mounted between them. I attached each end of the top of the one foot direction of the frame to the top of each respective post. This was my swivel point. I attached a leg from a lower point on the post up to the bottom of the frame to create the proper angle of the modules, such that the modules were perpendicular to the angle of the noon-day sun. The legs had other sets of holes to seasonally adjust the angle of the modules for the change in the position of the sun in different seasons.

The posts rose high enough out of the ground so that the bottoms of the modules were four feet above the ground. The snow cover on Bashan Hill is often three feet. I got 10 gauge (ga) wires and copper lugs from Wilson. I soldered the lugs on the ends of two wires for interconnects to wire the two positive terminals of the modules together and the two negative terminals together. I borrowed an impact drill from Wilson and drilled a hole through the foundation wall into the cellar. I ran a 10 ga UF cable from the modules, down one post, and underground inside a one-inch polyethylene pipe into the cellar.

In the cellar I built a strong table to hold the 280 pounds of batteries above the concrete floor. I made interconnects for the batteries from short pieces of 4 ga wire I had salvaged from Bill's junk pile. I stripped each end of the cables and attached bolt-on heavy-duty copper lugs. I installed anti-corrosion rings on each of the battery terminals. The four 6V batteries were grouped and wired such that two pairs of 6V batteries effectively became two large 12V batteries. These two batteries were paralleled to make a 12V battery bank.

I now had the modules wired and the supply line in the cellar. The four 6V batteries were physically and electrically installed to create a 12V battery bank. Today, a normal installation would always have a charge controller between the batteries and the solar electric modules to disconnect the modules whenever the batteries are fully charged. The *PV Network News* had devoted much verbiage on how to avoid the expense of a $100 charge controller. It was certain that the two modules that I had could never produce a high enough current to damage the battery bank that I had installed. When the battery bank was fully charged, the modules would overcharge the batteries only at a trickle charge rate. This would only result in a little more water being electrolized.

In the absence of a charge controller, I installed a Schottky diode to prevent a back flow of electricity to the modules during the night. I also installed an old recycled wall switch so that I would be able to manually disconnect the modules when I was sure that the batteries were fully charged. I specifically used an old switch because it was a snapping-type instead of a silent modern one. The snap of the switch meant it was rated for both DC and AC electricity instead of only AC electricity. A silent-type, AC-only, switch would quickly fail. At the battery bank, I installed a simple in-line car fuse in the circuit. This was a must for safety. A short in the wires between the battery bank and the modules would allow the batteries to short out. Some of these wires were inside the house where the shorting wires could start a fire in the absence of a fuse.

I had ordered a small auto fuse box and some other assorted parts from an alternative energy company in California. I did not want to wait for the order to arrive before I used my new solar electric system. This was actually a wise choice. My order arrived late with only a few of the items. The rest of the items were back ordered. The goods finally dribbled in over the next month. Many of the items were the cheapest RV quality at the most expensive price. Ultimately, this experience convinced me to use Bill Wilson's experience and my ingenuity to avoid that part of the alternative energy world.

I installed one standard duplex receptacle in my living room. It was improper to use a 120V receptacle in a 12V circuit. I color coded a brown receptacle with a white cover plate. I felt I could not miss that odd color scheme and accidentally plug a 120VAC appliance into the 12VDC system. I sent a 12 ga wire back to the battery bank. Since I had not yet received my fuse box from the mail-order company, I installed a single in-line fuse in the circuit at the battery bank. Upstairs in the living room, I hung a pull-string socket in a Chinese paper globe. In the socket I installed a 12V 25 watt bulb with the standard Edison base of a regular household bulb. I plugged the light into my color-coded 12V receptacle.

I double-checked all my wiring and fuses. It was now the end of the day. I went to the living room to enjoy my first solar electric powered light. This was the first time I had had light without the noise of the generator. It was truly magic. I felt for the first week that the generator was going to start and roar its roar whenever I pulled the chain and lighted my light. I was ecstatic over my new life. Bob and I would now get together and gossip about what we could do after dark with light in the house.

A SOLAR ELECTRIC SYSTEM produces a certain amount of electrical energy based on the number of modules and the size of each module. This electricity is used to recharge the batteries after the nightly loads of the house have lowered the batteries below 100% charge. One must monitor the batteries to make sure that, on an average, the house is not using more electricity than the modules are producing. If the house does use more than the modules

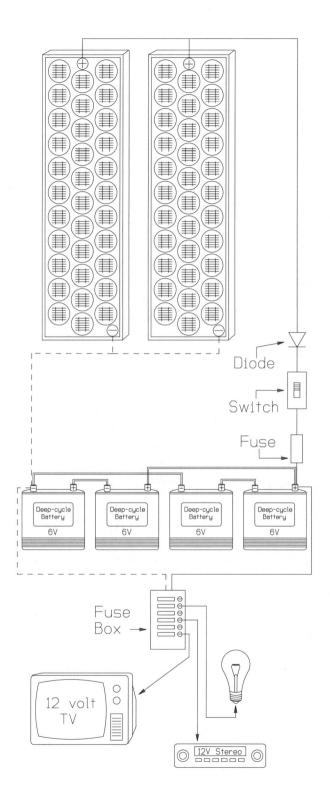

Diode

Switch

Fuse

Deep-cycle Battery 6V

Deep-cycle Battery 6V

Deep-cycle Battery 6V

Deep-cycle Battery 6V

Fuse Box →

12 volt TV

12V Stereo

My first solar electric system.

produce, then the batteries will gradually decrease in charge until they are ultimately dead. The simplest method of monitoring the state of charge of a battery bank is to monitor its voltage. A full battery bank has a rest voltage of about 12.6 volts; a dead battery bank has a rest voltage of about 11 volts. There are charts available that approximately relate the resting voltage of a battery bank to the percentage of electrical energy that is in the battery bank.

This small solar electric system had no meters. There was no ammeter to show how many amps were charging the battery bank. Joel Davidson had recommended testing the modules upon arrival to be sure they were working. Considering his years of experience with solar modules I am sure he knew that the solar modules would be working. Joel wanted people to be certain that they worked, so that they would not assume that their bad wiring was a bad module. I had used a multimeter (combination ammeter and voltmeter) to test the modules. After I installed the solar electric system I tested the batteries and the light circuit. The system was installed and working. It was now approaching the middle of November, the hardest part of the year to match the production of the modules to the loads of the house. Most of my loads would be lighting. The early nights of winter meant I would need the most hours of electric light at the time of the year when the least sunlight made the least electricity.

I started testing the voltage of the battery bank each night when the battery voltage was at rest. Each evening the voltage was less. I was using only one light, and there were sunny days. I then started testing the voltage at the end of a sunny day before I had used my first light for the evening. The voltage had not increased. I could find nothing wrong with my wiring. I was sure that the modules were working. Finally, I decided to test the open circuit voltage of the modules. I shut off the switch between the modules and the battery bank to test the modules at the two wires on the module side of the switch. The modules were working fine. I tested the voltage on the battery side of the switch. The voltage of the batteries did not increase when I switched the switch on and theoretically connected the modules to the batteries. Somehow the electricity from the modules was not reaching the batteries. I took the switch out of the circuit and connected the modules directly to the batteries. The problem, of course, was the switch. I had spent $1,100 on equipment, and the system was not working because of an old wall switch. I had been cutting back my use each day thinking that solar electricity did not really work. And all because of a bad switch.

With a new switch in place the modules recharged the batteries to full capacity in a single bright sunny day. I eventually received my auto fuse box from California. I had chosen the more expensive $12 model because the six-terminal fuse strip was enclosed in a metal box. I hated the idea of exposed live wires. Many alternative energy systems were actually a snaggle of exposed wires. Low voltage DC electricity will not cross the insulating barrier of human skin. However, it will arc and short if a metal tool touches exposed terminals.

The enclosed fuse box had two major problems. It was too small to hold six heavy wires, and it did not have enough wire access holes in it. I borrowed some of Wilson's tools and added holes that would accommodate standard connectors for house wiring. The small interior could not be enlarged. I just worked extra hard folding the large wires in the small space. The box was designed for the small glass auto fuses. These fuses were perfectly safe for this application.

The whole house was rough wired to the boxes at this point. Most of the downstairs walls and some of the upstairs walls were closed in. I decided to wire the house additionally for 12VDC. I found a way to get at least one receptacle or light fixture to each room of the house. This provided a minimum of one 12VDC light in each room. Each circuit had to be custom designed. Long runs at 120VAC with many receptacles per circuit are fine. Long runs at 12VDC with many receptacles will result in a voltage drop, which will result in dim incandescent lights, slower turning motors, and efficiency loss. I planned each circuit according to the wire size, the potential amperage draw of all the prospective appliances on the circuit, and the wire length of the circuit. When the job was complete, I had six relatively short 12VDC circuits with only one to three receptacles per circuit. Most of the receptacles were in just the right place for optimal use. Contrarily, my standard 120VAC house wiring had long wire runs with many receptacles such that you could always find one in the right place.

During the first year of my solar electric system my primary loads were lights. I was only living in the downstairs of my house. I lived alone. A large living room, a good sized kitchen, a bathroom, and a bedroom were all that I needed. I had one light in each room. In the winter I was very careful to light only the room where I was sitting. This meant I was using just one 25 watt 12V bulb. In the living room I had a 12V black-and-white television. The TV was rated to use 16 watts. Under actual test conditions it only used 12 watts. I also had a 12V stereo receiver and turntable. Under normal use the stereo consumed 5 watts, and the turntable consumed 2 watts.

The 25 watt 12V incandescent bulbs I used were sold for RV use. These bulbs looked the same as a standard household light bulb and had a standard Edison screw base. They were expensive, but convenient because they could be used in standard lamps or fixtures. Inside the bulb, the tungsten filament was a large wire size to utilize 12V electricity instead of the smaller tungsten wire size in a standard 120V light bulb. A 25 watt 120V bulb drew about .2 amps. The 25 watt 12V bulb drew 2 amps. A tungsten filament is more efficient at higher amperages. This meant more light and less heat per watt. Practically, this meant that a 25 watt 12V bulb yielded twice the light of a 25 watt 120V bulb. Or, my 25 watt bulbs were yielding an amount of light similar to a 50 watt standard household bulb.

I used less loads that first year than I could have. I was not accustomed to the average amount of the solar electricity that was available. I was hesitant to use much electricity when we had had some cloudy weather. I was afraid that the sun would not shine for days, and that the batteries would run down to a level where it would be difficult to return them to full charge. In retrospect, I could have used more electricity every day. The large battery bank and the average of the sunny days would have kept recharging the battery bank to full.

I had the opposite problem when summer arrived. The days were long. I had a nominal need for lights. My other loads were minor. There was no charge controller, and the batteries were overcharging everyday. The rate of overcharge was low, so no harm was done. However, I listened to the batteries bubble, and I had to replenish the lost water. I started leaving a light on all day to use the extra electricity. I had some misconception that if I disconnected the modules and left them in the sun, the electricity would somehow back up in them and harm them. Eventually, I learned that the modules could be disconnected. At that point I manually disconnected them on selected days.

I was more than happy with my small solar electric system. My greatest fun came when my friend, Dave Whitcomb, called to ask if I had power at my house. He had lost his power unexpectedly and wanted to know how widespread the outage was. I told him that I had power. At the close of the conversation I suggested that I probably was not the best person to consult on this subject. Lea and I still take pleasure playing this same joke about once every year.

The only discouraging part of the solar electric system was the installation. It was much too difficult considering the simplicity of the actual system. I had not had adequate directions. I had only been able to purchase the major components. I had used more time finding the small parts than I had used actually installing the equipment. In later years I looked back at these frustrations and tried to change the process for others who were following with their solar electric systems. When I started my own solar electric business, I tried to supply all the parts that were needed. I arranged components into kits. A customer could buy all the pieces for the complete system if he so chose. Additionally, I supplied diagrams and directions for each kit. As time went on, I wrote first *Solar Electricity for the Remote Site Home* and later *The Solar Electric Independent Home Book*. The 13,000 copies of these books we sold aided many people in buying and installing their own solar electric systems.

CHAPTER 18

Inverters

A N INVERTER is a power conditioning device that changes low voltage DC (direct current) into higher voltage AC (alternating current). Direct current is the type of electricity that is supplied by a battery. The electrons that flow through the wire in a circuit flow constantly in one direction. Alternating current is the type of electricity supplied by a rotating alternator, or generator. This can be a portable generator or a massive generator at the power company. Because of the design of the windings inside the generator, the electricity flows in one direction, then stops and flows in the opposite direction. One forward flow followed by one reverse flow constitutes one cycle of alternating current. The conventional grid electricity produces 60 of these cycles per second.

Alternative energy homes produce DC electricity to charge a battery bank. This battery bank can only supply DC electricity to the loads. As an alternative energy system grows beyond the smallest of systems and approaches systems that are meant to power the equivalent needs of a conventional home, it becomes necessary to add an inverter to the system to change the low voltage DC electricity supplied by the battery bank to the 120VAC electricity of a conventional home.

The oldest and simplest inverters were called motor generators. These consisted of a large DC motor whose rotating shaft was coupled to the shaft of a 120VAC generator. This was similar to a portable generator, only the gasoline engine that turned the AC generator was replaced by a DC motor. The DC motor was connected to a low voltage battery bank. The low voltage DC electricity turned the large DC motor, which turned the 120VAC generator to produce 120VAC power for household loads. These devices were reliable but extremely inefficient.

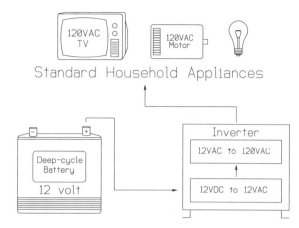

Standard Household Appliances

Inverter
12VAC to 120VAC
12VDC to 12VAC

Deep-cycle Battery
12 volt

An inverter changes low voltage DC electricity to conventional 120VAC electricity.

The modern inverter is an electronic black box. Internally, the inverter changes the low voltage DC electricity into low voltage AC electricity. A transformer then converts the low voltage AC electricity to 120VAC electricity. The simplest inverters make a very crude square-wave type of AC electricity. The best inverters make a true sine-wave form of AC electricity that is identical to the electricity from the power company. Most alternative energy homes utilize a quasi–sine wave inverter that is a good approximation of the sine-wave power from the utility company.

When I first started living on Bashan Hill, there was no good alternative to the AC electricity of the power company. Most alternative homes found ways to utilize DC for the majority of their loads. RV and marine DC appliances did a few jobs. The more advanced mechanics and electricians converted some appliances such as washing machines to low voltage DC. The majority of homes did without. Hand tools were used when appropriate. The loads that absolutely had to be powered by 120 VAC were powered by a gasoline, propane, or diesel generator.

In 1982, when I installed my first solar electric system, electronic inverters were available, but none of them was very good for my alternative energy home. The most popular one was a Tripp-Lite inverter, sold by Sears and by mail-order auto-supply companies such as J.C. Whitney. The largest power output from a Tripp-Lite was 500 watts. These inverters only produced a square wave. They were not good at running loads other than resistive loads such as light bulbs or heating elements. At best, these inverters were 60 percent efficient.

An electric motor consumes a large amount of electrical energy to boost it from off up to full rotational speed. This is referred to as surge. The amount of surge that is required is dependent on the design of the motor. An average example would be a 500 watt motor that takes a 1500 watt surge to start it. The surge required is three times greater than the power needed to only run the motor. A small Tripp-Lite inverter had no ability to supply a motor with a surge beyond the rating of the inverter. It could theoretically run a 500 watt motor, but it could not supply the 1500 watt surge required to start it. That meant that a 500 watt Tripp-Lite inverter could only start a 165 watt motor that needed a three-times surge of 495 watts. Larger inverters were rated for their running watts and their surging watts. A larger inverter might be rated to run a 1000 watt motor, because it was a 1000 watt inverter that could surge to 3000 watts for a short period of time.

The most reliable inverters available were those designed for military use. Unfortunately, these inverters were way out of the price range of the alternative energy homeowner. One of these inverters could cost as much as a whole large wind system. Best Inverter Company was selling inverters that many catalogs recommended for use with wind systems. These inverters were expensive, but not exorbitantly expensive. The Best inverters were reputed to be somewhat unreliable, as were most other inverters in their class. I do not mean that they were junk. Their design was not advanced enough to satisfy all kinds of loads under all conditions. When the inverter attempted to power the wrong combination of loads at the wrong time, it could not sufficiently protect itself from overload and failure.

The most efficient inverters were 80 percent efficient. This maximum efficiency was reached for large loads, above 50 percent of its rated output. For smaller loads the efficiency was much lower. For the very small loads of 100 watts or less the efficiency could be as low as 50 percent. When one of these inverters was turned on with no load being powered, the inverter used a minimum of 40 watts to power its internal circuitry. This was such a large resting load for a small alternative energy system that the inverter really had to be turned off except to power a specific load. Some inverter companies were offering a load-sensing option. These inverters used only four to eight watts while idling and waiting for a load to be applied. The inverter sensed a load and turned itself on to power the load. When the load was turned off, the inverter turned itself off, and again idled, only drawing four to eight watts. Even only four watts of standby power was really unacceptable for a solar electric system. Four watts for 24 hours per day was 100 watt-hours per day. This was equal to the winter output of one solar electric module.

After one winter with my solar electric system, I was convinced that it was to be my alternative energy power source. I was willing to invest more money and expand the system. However, I could not do so significantly until I could transfer loads from the generator to the solar electric system. I needed a large solar electric system to power a large inverter, which would in turn power my AC loads, thus making my generator obsolete. Any inverter that I would use would have to be reliable. I could not be without water for weeks while an inverter was being serviced. The inverters I had researched were not reliable enough, not efficient enough, and not really designed for use in an alternative energy home.

I received my first literature describing the new Heart Interface inverters in the fall of 1983. These inverters, designed for use in an alternative energy home, reached a suitable level of efficiency powering a 50 watt load. They were 90 percent efficient for larger loads. Every unit had a standard load sensing feature. The inverter only consumed $1/3$ of a watt of power when there were no loads in the house! The inverters were designed to be on at all times, ready to instantly power up when they sensed loads. Because the Heart Interface inverters were efficient even at low loads, there was the potential for powering a whole alternative energy home on 120VAC electricity. Efficient fluorescent lights could be powered by the inverter, eliminating the need for special circuits for efficient low-voltage DC lighting.

The introduction of the Heart Interface inverters instantly advanced the alternative energy world. The inverters were well made. Everyone could see the advantages of a piece of equipment that was designed specifically for use in a solar electric system. I called Heart Interface to inquire about becoming a dealer of these inverters. I was in the process of becoming a dealer of solar electric modules and batteries. Heart Interface's one requirement for dealer status was an initial order of two or more units. The first inverter would, of course, be for my own home. I solved the problem of the second inverter by offering an inverter at dealer cost to my friend Richard.

My first Heart Interface inverter which currently is powering another independent home in Worthington.

I wanted an inverter that would power my deep-well pump. A submersible pump is a difficult load to power because the motor is always starting under the load of the water pressure already in the water system. The motor requires a surge of electricity to not only boost the motor from rest to full speed, but from rest under load to full speed. The pump itself drew 1,000 watts. Instead of the normal 3,000 watt surge of a conventional motor, this pump required a higher surge. The exact rating of surge that the pump needed really depended on the time interval over which the surge was measured. The inverter's rating of surge capability was also dependent upon the time interval over which it was measured. Unfortunately there was no industry standard for the surge rating of the pump or the surge rating of the inverter.

I made many calls to Heart Interface to address the problem of the water pump load. They were confident that Heart Interface inverters in the field were powering similar loads. If the inverter failed to power the pump, they would be happy to have it returned. With the problem load of the water pump solved, I had the future design of my solar electric system solved. All other loads in my home could either be powered by the inverter, or they could eventually be eliminated. If the new system performed reliably, I would be able to sell my generator and invest the money in a larger solar electric system.

The 2,500 watt inverter that I purchased from Heart Interface was a 24V inverter. I had to convert my 12V solar electric system to a 24V solar electric system. A solar electric system is modular. All that I really had to do was reconfigure the wiring of my array of solar electric modules and the wiring of my batteries. The 12V circuits in the house would actually perform better at 24 volts. The real change involved the replacement of my 12 VDC lights and appliances with 24VDC lights and appliances. This would be well worth the effort to be able to power the well pump without the use of a generator.

THAT FIRST Heart Interface inverter powered the deep-well pump as planned. The inverter worked well for six months, until it failed for no obvious reason. I came home one night from graduate school in New Hampshire and turned on the water pump, only to watch a column of smoke slowly spiral up out of the top vent of the inverter. The inverter was still under warranty. I was a dealer. It would be fixed. The problem lay in the length of time it took to get repaired. The inverter was too heavy to ship via UPS. The freight trip to Washington state was slow. I lived without water from my water pump for three weeks while the inverter was being fixed.

By the time my Heart Interface inverter had failed, I had sold several more of them. Soon these too began to fail. Some failed because they were susceptible to transient electricity from lightning strikes. Others failed because they could not adequately self-protect themselves under all load conditions. It seemed that the wrong load or the wrong combination of loads overloaded the inverter before it could shut itself down. The overloaded Darlington transistors then shorted and smoked. My customers were not happy owning an expensive inverter that failed and took weeks to be repaired, even if the repair was free under warranty.

I personally went through similar headaches with other manufacturers' inverters. One inverter continually blew its load-demand circuit, while another inverter overloaded all its Darlington transistors as the Heart Interface had done. I was beginning to feel that inverters were not meant to be reliable.

IN 1986, I met Steve Johnson at a large renewable energy conference in California. Steve was the top man and partner at Trace Engineering Co., the maker of the new Trace inverters. These inverters replaced Darlington transistors with FET transistors. The Trace inverters were well self-protected but also had similar features to the original Heart Interface inverters. I ordered my first Trace inverter at the show. By the time I arrived home there was a message at Fowler Solar Electric Inc. from Steve at Trace. He wanted me to consider being a distributor of Trace inverters instead of being just a dealer.

As time passed I became good friends with the partners and employees at Trace Engineering. I realized I had worked with some of these people in the past. Steve Johnson had been a principal person at Heart Interface. He and others had left that company when my inverters were failing. They formed Trace Engineering with the firm conviction that they did not want to manufacture

A Trace 2,500 watt
inverter. Photo provided
by Trace Engineering
Company, Inc.

inverters that were unreliable, nor did they want to listen to angry customers.

The first Trace inverter that I installed in my own home was rated at 2,000 watts. It actually outperformed other inverters that had slightly higher ratings. The Trace inverter reached a critical power output that was extremely important to my home. The inverter could run both the deep-well pump and the washing machine at the same time. It could also handle the starting surge of either appliance at the same time the other appliance was operating. With earlier inverters I first had to run the pump to fill my water storage tank. Then I had to turn the pump off and start the washer. This ensured that the pump did not come on when the washer was running and send the inverter into overload.

Trace inverters improved the standby option that was available on the Heart Interface inverters. The standby option consisted of a transfer switch, a battery charger, and a sensor to sense 120VAC power from another source. In an alternative energy home, the other source of 120VAC power was a generator. Under normal operation, the inverter inverted the low-voltage DC power from the battery bank into 120VAC power for the home. When the generator was started, the inverter sensed the source of power. The inverter then waited ten seconds for the generator to reach full speed. The inverter then switched all loads in the house from the inverter straight through to the generator. At the same time, the inverter began to effectively

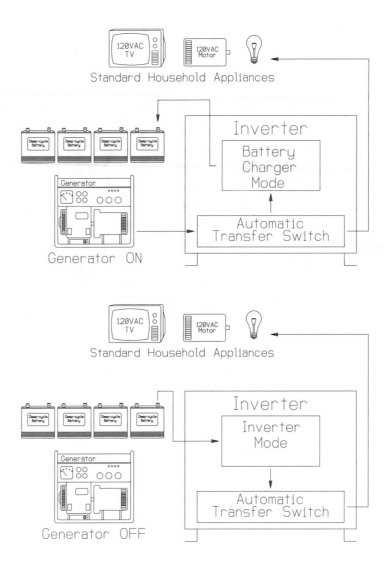

Standard Household Appliances

An inverter with a stand-by option installed to charge a battery bank with a standby generator.

work backwards as a battery charger. Finally, the generator was powering the loads in the home and charging the battery bank. When the generator was turned off, the inverter automatically returned all loads to the inverter and the battery bank.

Trace improved the standby option by providing a programmable battery charger. One could customize the operation of this charger to the size of the battery bank and the size of the generator. I had no generator at this time. Nevertheless, I bought the battery charger option for my Trace inverter. I found this to be great insurance. If the battery bank ever became hopelessly discharged, I could borrow or rent a generator for a day to recharge it.

Future 24V model Trace inverters increased in rated output from 2,000 watts to 2,500 watts, and later to 2,600 watts. Because I owned a solar electric equipment business, I always had the latest model. The 2,500 watt model was a big increase in size for my home. There was no combination of common loads in my house that would overload this new, larger inverter. I could use my Sears radial-arm saw for the first time since I had sold my generator. This was the most difficult motor I had to start on an inverter. The motor required an extremely high surge to start it.

LIGHTNING was my biggest headache concerning inverters. As they became reliable in design, it became clear that most of the few failures of my customers' inverters were actually from transient voltage surges from lightning. The lightning did not strike directly and destroy appliances and the inverter. The lightning struck near the house and induced a high-voltage surge onto the system wiring that effectively erased the memorized data on a chip on the circuit board.

My land and the area on Bashan Hill is very susceptible to lightning. At first I disconnected all circuits from the outside of the house to eliminate my susceptibility to lightning surges. I added lightning suppressors to all incoming and outgoing circuits. Finally, I became quite sure that the greatest susceptibility to lightning for my inverter came from my deep well pump. Lighting striking near the well casing induced a high-voltage surge on the well pump wires that carried the surge into the house. I improved the grounding of my home, the plumbing, and the well pump. I added multiple lighting arrestors to the well-pump circuit. Finally, I added a large switch box to the pump circuit that could disconnect all wires to the pump. This switch is installed inside the cellar stairway. When lightning threatens, I open the cellar door and disconnect the well pump. When I am not home, I leave the lightning arrestors to do their job alone.

An inverter has become just another part of the way of life at our alternative energy home. For the past eight years, we have lived free of any failures. The idiosyncrasies of quasi–sine wave electricity are relatively unapparent at this point. We have chosen our appliances to be compatible with our inverter. I no longer dread lightning. The inverter is adequately protected by its own design and also by my design of lightning protection for the whole house. The first three years of problems were a worthwhile part of the total evolution of the alternative energy system.

CHAPTER 19

My Second Solar Electric System

B Y THE FALL of 1983 I had lived with my two-module solar electric system for almost a year. I was certain that this was the proper energy source for my alternative energy home. My thoughts moved to expansion. At the time, most of us solar electric users were pulled in two different directions when we thought of expanding our solar electric module array. On one side, we were all waiting for the magic breakthrough in technology that would dramatically reduce the price of the solar electric cells used in the modules. On the other side, we wanted to buy all the modules that we could before the federal tax credits for renewable energy systems ran out.

As fall proceeded and the work days shortened with the earlier setting of the sun, I began to plan for an increase in my system before winter. At the same time I also began to consider using my knowledge to sell solar electric equipment as a part-time business. Joel Davidson had left rural Arkansas for Los Angeles, of all places. Bill Lamb had convinced Joel that he could most impact solar electric proliferation as a sales manager of solar electric equipment at Wm. Lamb Corporation. The advertisements that Wm. Lamb Corporation was running in the alternative energy magazines finished with a small line at the bottom, "Dealer Inquiries Welcome." Joel and I spoke for an hour on the telephone. Joel had two objectives. First he wanted to make sure that I knew what I was doing. Secondly, he wanted to get as many solar electric modules out into the real world as possible. As soon as he found my expertise was sound, he made me a dealer.

I installed a new frame and six modules on the existing posts that had held my previous two-module frame.

As in all dealer agreements, the unit price of modules did not become appealing without a significant quantity order. I could not just order two more modules for my own solar electric system and lay dormant waiting for my first customer. I made a few phone calls. I offered modules at wholesale prices to my friends in order to create a quantity order. Bob Cook and Richard Mansfield each wanted two modules. To their orders I added an order of four modules for my own home. After a day of worry, I chose to order an additional four modules for my new business venture. I felt the $1,100 investment in stock for my new business would serve as a desperate act of commitment. I could not afford to get side-tracked. I would be in the solar electric business no matter what. After all, this seemed to be where fate had been pushing me all along.

After I had ordered modules from Joel, I received a retail flyer introducing the new Heart Interface inverter. This new inverter seemed to have the functions that were necessary for use in an alternative energy home. I had always planned to incorporate a large inverter into my alternative energy system. I would never be able to retire the generator without an inverter as a replacement producer of my much needed 120VAC power. I could not convert my well pump to low voltage DC. Many other appliances would be impossible to run on low voltage DC, or they would be impractical to convert.

Heart Interface made a 1,000 watt 12V inverter and a 2,500 watt 24V inverter. The smaller inverter was definitely too small to power my deep well pump. The 24V, 2,500-watt model appeared to be large enough to satisfy the extreme surge necessary to start my 1/3 horsepower pump. I explained to the sales manager at Heart Interface that I was a dealer of solar electric modules and a dealer of Exide batteries. He offered me dealer pricing of 30% off list on Heart Interface inverters with an initial order of two or more.

At this point I was playing an amicable shell game with the distributors of solar electric equipment. I was telling each manufacturer that I was a dealer of the other goods so I could be a dealer of theirs. I had the potential to be a good dealer, but realistically I was only a knowledgeable individual who wanted to expand my system at non-retail prices. I felt I was telling half truths. What I did not understand was that the manufacturers did not care. They were desperate to move their products. The majority of their dealers were located in California and parts of the southwest. The distributors were searching for dealers in the relatively uncharted territory of New England.

Now that I had my potential dealership of Heart Interface Inverters, I got back on the telephone to look for a prospective customer for my second inverter. Bob Cook was not interested. He would always need his generator because it was also his welder. He did not have any critical daily AC loads. He only needed AC power for tools. He had been my most likely candidate, and he had given me a definite no. My last hope was Richard. Richard surprised me with a quick yes for a 1,000 watt inverter, which listed for $900 with a dealer cost of $560. My 2,500 watt inverter listed for $1,300, with a dealer cost of $910. The true cost for my inverter would be less than $400 after I used my energy tax credits. This was a rather small gamble for an inverter and a start in a business that I would enjoy.

My system expansion plans were now complete. I would add four 35 watt modules to my two existing 33 watt modules. I would add two batteries to my four-battery, battery bank. I would install a charge controller and a 2,500 watt, 24V inverter. Up until this time I had had a 12V solar electric system. I would now need to rewire my modules to make a 24V array to charge a rewired 24V battery bank. The 12VDC fusing, wiring, and appliances would also have to be upgraded or converted to 24VDC.

The modules arrived in December, long before the inverter. They were all ordered at the same time, but Heart Interface could not keep up with its instant success. As a small and new dealer, I waited. My 2,500 watt inverter did not arrive until February. I had installed the other new components in my system long before my inverter arrived.

I REMOVED THE FRAME that held my two original modules. It was winter, and the ground was frozen. It was easiest to build a new frame to mount on the existing two supporting posts that were in the frozen ground. This new frame would accommodate the two original modules plus the four new modules. Working in my warm cellar, I built the frame, mounted the modules, and wired them as a 24V array. I enlisted Richard to help me carry the finished unit outside to be mounted on the posts. Even though I now had a six-module array, the existing underground wire to the house was still the proper size. The wire had a lower resistance at the new higher voltage of 24 volts.

I abandoned the "bubble gum and baling wire" approach of the first system and added a charge controller. Besides the addition of a controller, I wanted the wiring in this new installation to be similar in quality to my conventional 120VAC wiring. I ordered options for the controller. The most important option was a set of meters. The voltmeter displayed battery voltage so I could conveniently monitor the level of charge of the battery bank. The ammeter displayed the amps that were being produced by the modules. For the first time I could actually "see" my modules sending amps to the battery bank, and I could "see" the battery bank voltage rise as the battery bank was being charged. This controller also had a status light, which indicated that the batteries were charged. When the batteries reached a state of full charge, the controller decreased the current from the modules to a trickle charge.

I also ordered an array-fuse option and array-diversion option. The array-fuse option replaced the basic in-line auto fuse of my first system, with a fuse neatly mounted on the face panel of the controller. Initially, the removal of this fuse served as my disconnect for the array of modules. A few months later, Bill Wilson salvaged a used disconnect box that I added before the charge controller. The module array diversion was an option I had badly wanted. This option directed the output of the modules to another place, such as an additional battery bank, when the modules had completely charged my battery bank.

The greatest inefficiency of any home solar electric system is the loss of output from the modules once the charge controller has disconnected the modules from the fully charged battery bank. All the potential production of the sun shining on the modules is lost. The house cannot use the power at the time, and the storage system cannot store any additional electrical energy. This inefficiency was clear to me even with my first two-module system. I bought the array-diversion option as a solution. Unfortunately, I was never able to utilize the option. The method of array diversion was useless for me. It could not be used to charge additional batteries because there was no

I added four batteries to my original battery bank and rewired it from 12 volts to 24 volts.

charge controller in the diversion circuit to protect the second battery bank from overcharge and damage. The modules could not power an attic fan or motor load, because the output of my array was too large for such a load.

Before the inverter arrived in February, I had completely expanded the other parts of my solar electric system. One day, I was testing the voltage of each individual battery in my six-battery, battery bank. Each battery should have registered the same voltage, taking into account a .05V maximum variance. The voltage of the two new batteries was much lower than that of the four older batteries. I tested the batteries at different times and found that the two new batteries showed the same voltage when the battery bank was full and a lower voltage when the system was down. I realized I had made a silly error in the design of my battery bank. I really needed the battery bank to consist of multiple groups of four 6V batteries in series. Each group of four 6V batteries effectively would make a large 24V battery. In my haste to save money, I had purchased only two additional batteries. I had created a battery bank that could not stay in balance, because it had only six batteries. I really needed to wire my first four batteries in series to make a large 24V battery, then buy four more batteries to make the second large 24V battery. I made another trip to Boston for two more batteries to correct my error.

After two months of preparation and anticipation, the freight truck delivered my ninety-pound inverter. Under normal operation my 2,500 watt inverter would use a maximum of 100 amps from the 24V battery bank. Under motor-starting conditions, it could use several hundred amps. The efficiency of the inverter and its ability to deliver a large surge were dependent upon a lack of voltage drop between the battery bank and the inverter. I would need large, short inverter cables to prevent a voltage drop. With the aid of Wilson and his parts supply, I made a pair of four-foot-long "0"ga cables with heavy-duty bolt-on lugs for connection to the inverter and the battery bank. I placed the inverter close to the battery bank to accommodate the efficient, short cables.

The greatest danger of installing a new inverter is reverse wiring the positive and negative input cables from the battery bank. If I had reversed the two cables, the inverter would have self-destructed. I followed the old carpenter maxim: measure three times and only cut once. I triple-checked my wiring before connecting the last cable. When I attached the second cable, a large spark arced from the cable to the inverter terminal. I was sure I had blown the inverter by reversing the polarity of the wires. I checked a fourth time. The wires were correct. The second time there was no spark. I later learned that this was a normal and expected spark. The inverter takes a slug of current to fill a large capacitor. To this day, I still gasp whenever I hook up an inverter and endure that spark.

I turned the inverter on and plugged in a power drill. I pulled the trigger of the drill but nothing happened, so I released the trigger. I decided to try again. This time I held the trigger longer. The drill groaned and reached full speed over a span of a few seconds. I realized that during this delay, the inverter had sensed the load of the drill and turned itself on. It seemed like magic. I stood there and kept telling myself the drill was really being powered by those batteries, and that those batteries were powered by those modules outside that were sitting in the sun. This was the first time in several years that I had used a power tool without running out to start the generator first.

I got my circular saw. I tested it under heavy load by sawing a knotty 2x4. The inverter provided ample power. It was time for the mother of all loads, the water pump. I half expected the inverter to go up in a cloud of stinky electrical smoke. I plugged in the pump. The inverter made similar 60 cycle humming noises as it had with the previous loads. It seemed to be powering the pump as it had the saw. The pump was 200 feet away and down 80 feet. I walked across the cellar to see if the pressure gauge was showing an increasing pressure in the tank. The pump was working. I was on my way to a home powered by alternative energy.

The inverter had many idiosyncrasies. The first was the load-demand function. The first load applied to the inverter turned on the inverter. As the inverter powered up to full output capability it also had to power the existing load that had turned it on. This resulted in a motor groaning its way on. Wilson assured me that this was actually a very good way to start a motor. The voltage rose and the amperage was limited. It was not to be confused with a straining circular saw starting under load at the end of a long extension cord. In the latter case the saw draws high amperage that could burn out the motor. When the first load was a light bulb, the light would take a few seconds to go up to power. The switch would click, then the bulb would start to glow and increase to full intensity. Many years later, Dave, a customer from New Hampshire, came for the tour of my home. His wife said "oooh" when she saw the hall light slowly come on. She said she loved it because it would always remind her that it was powered by the sun. I realized that was actually how I had felt the first day.

THE INVERTER DID NOT PRODUCE the true sine-wave power of the power company or a portable generator. The quasi–sine wave output of the inverter resulted in some minor electrical interference, or electrical noise, in all circuits of the house. An incandescent bulb that was loose in its socket had a ringing hum. The controller for the deep-well pump in the basement made a humming noise while the pump worked. The inverter also produced RFI (radio frequency interference), which traveled through the air and spoiled my AM radio reception. These idiosyncrasies were only minor inconveniences to me.

Life with the new inverter became a testing ground for new loads. Some loads, such as my nephew's tube-powered Fender guitar amplifier, were not compatible with my inverter. The noise from the inverter power was picked up and amplified by the tube amplifier into a menacing feedback. The same thing happened with my old Dynaco stereo amplifier. The television had a small line of interference, but it was only visible if you were extremely close to the screen.

I began to experiment with AC lights. I wired the inverter to the input side of the 120VAC fuse box, or service panel, and thus fed all the 120VAC circuits in the house. Since the AC wiring in the house was extensive, I had the option of using an AC light almost anywhere. AC light placement was much more convenient than DC light placement, where I had only a few receptacles. I used standard 25 watt light bulbs in the hall ceiling fixtures. Light for these areas was strictly for making my way from room to room. These lights did not need to be exceptionally efficient because they were only on for a few minutes. I installed 60 watt bulbs in closet or cellar fixtures where I needed the strong light to find something, but here also the light would only be on for a few minutes.

I tested a standard fluorescent light in the cellar. This was a disaster. The ballast in the light made an obnoxiously loud 60 cycle hum because of the

Later that year, I increased my module array to ten modules and mounted them on the south side of the house. There was space for two more modules in the future.

quasi–sine wave power of the inverter. I bought a 22 watt GE Circlite. This consisted of an eight-inch circular fluorescent bulb mounted on an axis that contained the ballast and a standard Edison base for installation in a standard light fixture or screw base. The output from this 22 watt light was equivalent to the output of 60 watt incandescent bulb. This fluorescent light utilized a high frequency ballast that picked up none of the inverter noise. There were other oddities. The GE Circlite was a odd load. The inverter could not sense it as the first load and turn itself on. My solution was to turn on the switch to the Circlite in the kitchen, then turn on a hall light that would properly start the inverter, and then turn off the hall light. Once the inverter was on, the Circlite would light and keep the inverter on. The Circlite produced less than full illumination if it was the only load being powered by the inverter. I learned to use that light in conjunction with another load.

My battery bank had been rewired from the original 12V configuration to the 24V configuration of the new expanded solar electric system. All of the existing 12VDC circuits and fuses were now 24VDC circuits. I replaced all 12V bulbs with 24V bulbs. Unfortunately, I had no way to either convert or replace my 12V television and 12V stereo. These two appliances used very small amounts of electrical energy. I powered them on one circuit from a single 12V marine battery. This little 12VDC system was completely independent from the rest of the house. Once a month I recharged the marine battery at a friend's garage. This solution was a stopgap method. Within a year I had researched and purchased a $180 device that powered the 12VDC circuit to my 12V stereo and 12V television efficiently from my 24V battery bank.

I expanded my new 24VDC distribution system. The original 12VDC circuits were limited because of the amount of voltage drop in the 12VDC circuit wires. If I had wanted more 12V appliances, I would have had to use heavier wire and larger electrical boxes than are standard in a home. The conversion to 24VDC allowed me greater flexibility. A given wire size has four times less voltage drop for a given wattage appliance if the appliance runs on 24V instead of 12V. My change to 24VDC allowed me to add additional boxes to the old existing 12VDC circuits. I also added the longer circuits that were necessary to reach the upstairs rooms.

The inverter worked well. I was now committed to the elimination of my generator. Over the following months I found that I only needed to run the generator to maintain the charge of the battery that was used to start it. As the next winter approached, I decided to sell the generator and install four more modules on my solar electric array. I had never sold the extra four modules from my bulk order with Joel Davidson at Wm. Lamb Corporation. They were sitting in my upstairs hallway as a constant reminder of money not well invested. I had sold a few solar electric systems, but the module manufacturer had introduced a new, more powerful series of modules. I was always honest about the new modules. My customers never purchased my old stock.

In the late fall, I increased my array from six modules to ten modules. Once again I was making a new frame and rewiring the modules. This time Bob Cook welded some steel frames from scraps from his ironwork jobs. I mounted the new frames and modules on the south side of the house in a space between the top of the downstairs windows and the bottom of the upstairs windows. I ran new wires down the outside wall and into the cellar. I felt it was time to ground my solar electric system. There were now electrical code requirements for grounding a solar electric system that had higher-voltage components such as my inverter. I grounded the modules and frames to an outside ground rod. I also grounded the negative terminal of the battery bank to the same ground rod. I bonded this ground rod to the one that grounded the AC system of the house.

I sold my generator and repaid myself for some of the solar electric equipment I had purchased. The whole adventure was actually a legally rewarding economic experience. My solar electric system expansion of an inverter, a charge controller, eight additional modules, and parts had cost about $3,600. I received about $2,160 in state and federal tax credits. Thus my real cost was only $1,440. I sold my generator for $1,450. With the aid of the government's incentives, I had effectively traded my generator for the expansion of my solar electric system.

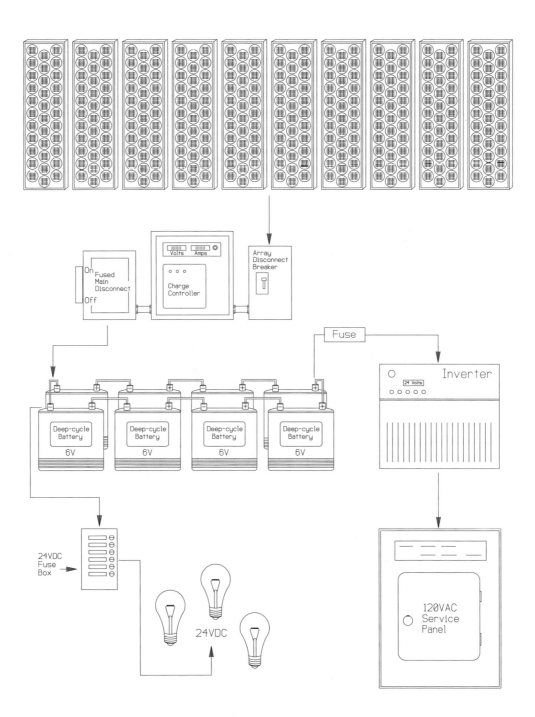

**My expanded
ten-module, 24V solar
electric system.**

CHAPTER 20

My Third Solar Electric System

MY TEN-MODULE solar electric system was more than adequate for me and my home. I lived alone with my dog, Rolo. My needs were modest. Being one person, I never needed to light more than one room at a time. I washed one person's dishes and took one person's showers. Most of the construction on the house did not take place during the low-sun period of winter. I had all the power I needed for my power tools, even after the generator was sold. I could have been happy for many years with my system. However, I had been bitten by the solar electric bug.

In the fall of 1985, I once again expanded my solar electric system. I replaced my ten-module 345 peak-watt system with a 24 module, 800 peak-watt system. While following a Masters Program in Environmental Studies, I realized my true desire was to be a leader in the solar electric field, rather than an entry level bureaucrat in the state environmental system. I wanted to change the world. I hoped to change the world incrementally with each new solar electric system that was installed as a result of my knowledge or my example. I was a product of the idealistic sixties. I did not have the bravado to proselytize the environmental cause. I did feel that I had the resolve to happily live my life by example, and the faith that my example might be of help to others.

Graduate studies at Antioch New England were flexible and innovative. I had second thoughts about graduate school after my first semester. My return to school had released me from my doldrums of working as an electrician for Wilson and living hand to mouth. I was growing suspicious of the lack of jobs in the environmental field. My solution was to accelerate my studies and complete the coursework early, before I quit the program.

For the summer, I arranged a six-credit independent study to write a how-to book for solar electric independent homes. Writing the book would require more work than completing six credits of classes. As my part-time business and contacts in the solar electric world had grown, I became painfully aware that people needed help to properly design and install a solar electric system. The people in the field who had the expertise had full-time work schedules and no additional time, a lack of writing skills, or both. I was not sure if I could write, and I was not sure if I had the knowledge. I did have the dream and the commitment.

In the middle of the book project, I attended a large three-day conference at MIT on solar electricity. The conference fee was expensive, but I attended hoping to fill the many holes in my knowledge. The speakers included many leaders in the field about whom I had only read. Karin Cook took care of Rolo. I had a free place to stay with a close friend, a professor at Tufts.

I arrived early at MIT the first day of the conference. I wandered around the reception area by myself, carefully procuring a copy of every free piece of printed material available. I did not know anyone. I was certain that everyone was an expert except for me. I looked like the hick in New York City who stares up at the skyscrapers. For whatever reason, a very nice, engaging woman made it her business to help me. She introduced herself as Marilyn. Marilyn asked all the right questions to draw me out of my shell and into the conference spirit. She seemed to know everyone, and she politely introduced me as her new friend to everyone who stopped to speak to her. She explained that her husband was one of the speakers.

I revealed to Marilyn that I sold solar electric components and that I lived in a solar electric powered home. Her natural and friendly manner defused my guard, and I modestly admitted that I was in the process of writing a how-to book for solar electric independent homes. She asked me who was publishing my book. I told her I planned to sell it to other solar electric dealers. I had unwittingly let my anemic little cat out of the bag. She explained that her husband was also writing a book on solar electricity that would be published by Rodale Press. She seemed to consider me an equal. I knew I was in over my head. From that point on, Marilyn introduced me to people as Jeff-who-lives-in-a-solar-electric-home-and-is-writing-a-book-on-solar-electricity.

I enjoyed the attention. On the other hand, I knew I might not even be capable of finishing my book because of lack of ability. The final slap from reality was delivered when Marilyn introduced me to her husband, Steve, who did not have the same last name. Her husband was in fact Steven Strong, a well-known solar architect and solar electric system designer.

My first book.

**Steven Strong.
Photo provided by
Steven Strong.**

I found Steve to be an imposing personality at that first meeting. Steve spoke in a very slow, deliberate manner. He was exceptionally serious. At first, I was not sure if he was irritated at having to meet me. Later in the day, when I heard Steve lecture, I could see that this was the way he spoke to everyone. He often stopped for a long, intense pause. There were never any *ahs* or *ands*, just painful pauses. I wondered if Steve stopped all action and concentrated during these pauses.

As the conference proceeded, my self esteem increased. I listened to the experts. They were professors, electrical engineers, and architects. I understood all of their lecture material. During years of my own study I had already learned most of the information that they shared. I actually felt superior to many of them on one small level: I was living with solar electricity every day.

Later in the conference, Steve Strong sought me out. He was very interested in my small solar electric business and my solar electric home. He loved solar electricity. He had made it his life's work and passion. A few months later, he and Marilyn came to my home to visit. He told me I had a life that he and Marilyn wished to have. I was living in a passive solar home powered by solar electricity in the peaceful countryside. Unfortunately, he made his living not from solar electricity, but from being an architect. Steve and Marilyn needed to stay in greater Boston where there were customers who could afford Steve's services.

Steve questioned me at the conference about my solar electric business. He wanted to know what modules I sold and where I purchased my goods. At the very end of this questioning, and in the most sincere non-salesman-like way, he stated that he was the northeast distributor of Mobil Solar solar electric modules. He would like to supply me with his product if I was interested. I asked him to send me all the appropriate information.

I arrived home energized from the conference. I was determined to write my book. I was also champing at the bit to expand my solar electric system. I wanted my home to be at the level of the homes featured in the lectures at the conference. At the time, I was purchasing my modules from Paul McClusky at Solar Electric Specialties Co. in California. Paul was encouraging me to expand my solar electric system to impress my customers. We talked of replacing my old-style modules with the latest model. Paul was also selling some second-grade (with minor defects) models of the old-style modules that were already on my house. On impulse, as a direct result of Paul's sales pitch, I ordered twelve second-grade modules on the Thursday night before I left for several business meetings in Boston.

On the trip to Boston, I had recurring second thoughts concerning the defects of the modules I had ordered. Friday morning I called Steve Strong to schedule our meeting. Steve also had reservations about the second-grade modules. He offered me some first-grade Mobil Solar modules at a discounted price. These modules were old stock because they were 6V modules. They were left over from a project. I would have to externally wire groups of four modules in series to reach the 24 volts that were needed in my system. I called Paul McClusky to cancel my order.

The nicest characteristic of the Mobil modules was that they would fit perfectly in the space on the south side of my home above the downstairs windows and under the upstairs windows. These modules were only 36 inches tall. They would fit in this space in an upright orientation. The Mobil modules were also a rich blue. They were the prettiest modules I had ever seen.

I drove to Steve's home that afternoon. I purchased sixteen 33 watt modules for my home. Steve and I visited during the ride to pick up the modules at the warehouse. We ate dinner with Marilyn. Steve and Marilyn and I would become good friends. Marilyn had been open and friendly from the first time she introduced herself. Steve was slower to open up, as I was, but by the end of dinner we had both dropped our business postures in favor of our personal ways.

Two weekends later, I received a call from Steve. He was calling from Jimmy Molyneux's real estate office in Worthington. He and Marilyn wanted to come and visit. At our last meeting they had confessed that they wished they could move to the country to build their dream solar house. I had given them the number of my good friend and real estate agent in Worthington. We visited and dined in nearby Northampton. It was a great social time. I hoped that Steve and Marilyn would buy land and build their dream house in my backyard.

At my house Steve revealed that he had had to purchase the whole lot of 26 6V Mobil modules in order to fill my order of sixteen. He also confessed that he had brought the last ten modules along in his car in case he could talk me into buying them. After I had returned home the week before with my first sixteen modules, I had realized that 24 modules would nicely fill the whole space on the south side of my house. I bought the ten modules Steve brought in his trunk. I could sell two modules through my business and add the other eight to my sixteen modules. I wanted to have a large solar electric system as a model for my customers, and I wanted to have an architecturally attractive installation. The 24 modules fit so well in the space, one would think they were an integral part of the greater house design. My house and 24 modules were subsequently featured in Steven Strong's book, *The Solar Electric House*, and Joel Davidson's book, *The New Solar Electric Home*.

IN THE EARLY YEARS of solar electricity, mounting structures were expensive, and too often were designed such that one design fit all installations. Most of the mounting structures oriented the modules' long dimensions from east to west. A four-module frame stacked four modules, one above the other. This was a bad design for snow climates like New England. The snow caught the horizontal edge of each module as it attempted to slump off the array. Modules installed in snow climates needed to be oriented such that the shorter dimension was heading east to west and the longer direction angled upward. In this orientation, slumping snow only needed to slide over the bottom edge of each module.

While I wrote my book and installed my 24 new modules, I worked 20 hours a week at the local Hilltown Community Development Corporation. My good friend and co-worker was a scavenger like me. Trish excitedly told me she had just been to the Chesterfield dump, where many old army cots had been discarded by the local Boy Scout Camp. Trish and I left for lunch early. We loaded ten angle-iron cots into the back of my Ford Bronco II before any town official could stop us. The two-inch by two-inch angle iron of these cots became the six mounting frames for my 24 new modules. They are still installed on my house.

To mount the 24 modules, I needed to drill 132 holes $5/16$" in diameter, and make 24 cuts in the angle iron. This turned out to be an enormous undertaking. I was burning up saber-saw blades and drill bits. These cots were authentic army issue. The angle iron was not standard strength steel but some sort of extra strength steel that must have been specified by the military. I eventually purchased extra-heavy-duty blades and drill bits to finish the project.

My earlier module mounts had been adjustable to achieve the best angle of insolation for the four seasons of the year. The majority of the homes featured at the MIT conference had module arrays that were not adjustable. These fixed-mount arrays were normally mounted co-planar to the roof; that is, on the roof in a parallel plane to the roof. This was a good average approach. In the southern latitudes the roof pitches tended to be flat, while in the northern latitudes the pitches were steep because of snow loading. My roof pitch was 45°. This angle coincided with the optimum angle of insolation for my area in the fall or the spring. If I had mounted my modules at a 45° fixed angle I would have received nearly optimum output in the spring and the fall, good output in the winter when the sun was lower in the sky, and good output in the summer when the sun was higher in the sky. This 45° orientation would have given me the greatest yearly output.

I considered mounting the modules on a nonadjustable mount at the optimum winter angle. The first part of my theory was that I had to get the best output in winter when I obviously would have the greatest loads and the least amount of sun. This would be most important to me, since I had no backup generator. I also needed the winter adjustment in the winter because I had snow cover on the ground for most of the season. In previous winters I had monitored the reflected light off of the snow. The output of my modules had increased 20 to 25 percent on bright sunny winter days when there was snow on the ground. This reflected light would be optimally picked up by the modules mounted at a steep 60° winter angle. If I used the more average

45° angle optimal for spring and fall, the modules would not be able to harvest the snow reflected light in the winter.

I carefully studied the monthly insolation data for my area. I had data that calculated the amount of module output per month at different module angle orientations. Modules mounted at the optimum winter angle (approximately 57° for my area) would obviously produce the best output for the winter months. I found that modules mounted at the optimum winter angle for the fall and the spring months produced about 5 percent less output than if they were adjusted to the optimum spring or fall angle of 42°. Modules mounted at the optimum winter angle for the summer months produced about 18 percent less output than if they were adjusted to the optimum summer angle of 27°. In summary, I felt that a fixed-module array adjusted at the optimum winter angle would give me optimum output plus the enhanced output from snow reflection when I needed it most. In the spring and fall it would give me nearly optimum output. I was not worried about the less than optimum output of the three summer months. I had the least amount of loads in the summer. I knew from my experience with past systems that my batteries would be charged early in the morning on any sunny summer day. I was committed to mounting my modules at a fixed optimum winter angle.

I FEEL THAT THERE ARE TWO general ways to size a solar electric system in the Northeast. One way is to install a solar electric system that will provide all the power necessary in the worst winter months. This system will be large

The controls for my 24-module solar electric system.

and expensive, and will have twice as much output as necessary for the summer. Most of the excess summer power will be wasted. The other way is to install a system that will meet the power needs of the house for the spring and the fall. There will be extra power in the summer and not enough in the winter. The home must either cut back on loads in the winter or supplement the electrical output of the solar modules with a backup generator.

I planned my new solar electric system of 24 modules to meet nearly all my winter loads. I planned to have a larger battery bank to average out the uneven patterns of winter sun. I planned to conserve in the winter, not because the monthly average output would not meet my load demand, but because I might have a period of no sun for up to three weeks. I would conserve by reducing the number of lights I used as the no-sun period lengthened. I could also pile up the laundry during the last part of a no-sun period. These plans were for my future use, and for the future use of a household that I hoped would someday include a whole family. Fortunately, solar electric systems are modular. If my plans were in error, I would always be able to add more modules. At the time, the large system would only need to meet the loads of one person and a part-time business.

Each serial-numbered module that I purchased from Steve came with a computer printout of exact test results. Using this data, I sorted the modules by their amperage output. To obtain optimum module output, modules that are wired in series to create higher-voltage arrays should all have identical outputs. Most modules are actually plus or minus 10% of their rated output. One can crudely but effectively test modules on a clear day with a multimeter to sort them. These particular modules could be sorted according to actual test data.

I mounted four modules per frame. This unit weighed about 60 pounds. I had learned a few tricks since my first days of installing a solar electric system with no directions. I had solved the incompatibility problem of aluminum module frames and steel bolts and steel mounting frames. I had learned that stainless steel was compatible with both steel and aluminum. I bought stainless steel nuts and bolts in quantity from a large supplier. To solve the problem of the contact of the aluminum module frame and the steel mounting frame, I inserted a stainless steel washer on each bolt as a separator at these points of contact.

The junction boxes on the back of the new modules were far superior to the single terminals on the old modules I had just uninstalled. Each module came with two watertight connectors for entering and exiting wires. I was able to neatly wire these modules up to electrical code in a manner that would require minimum maintenance for the life of the installation. Each group of four modules was wired in series such that they effectively became one large 24V module. From each group of four modules I sent one set of two wires down the wall and into the house to a junction box. Each of the six

The 24 Mobil Solar modules installed on my house.

groups of four modules was wired in this manner. I liked this approach because I could test each group of modules at the junction box in the base-ment. I could quickly troubleshoot and isolate a problem without pulling wires while standing on a ladder in the dead of winter. At the junction box in the cellar, the six pairs of wires fed two large wires that carried the array out-put to a DC rated circuit-breaker box, then to the charge controller, next to a fused disconnect, and finally to the battery bank. I was happy with the wiring. The whole installation was far superior to my first installation. I had an installation that satisfied the electrical code. The wiring finally looked like top grade home wiring.

My new 24-module array was rated at 800 peak watts. This was roughly equivalent to sixteen of the more recent 50 watt modules. An adequate bat-tery bank would have been sixteen of the 6V 200 amp-hour Exide batteries that I was selling at the time. I chose to oversize my battery bank because I had no generator to supplement my solar electric system in the winter when the sun shone least. The 24V battery bank had to be increased by incre-ments of four batteries. I finally settled on 24 batteries. I wanted to start my battery bank balanced, with all new batteries of the exact same size. I sold my used batteries to my friend Richard and to my friend Chris, who was installing his first small system. These eight batteries actually lasted a com-bined total of ten years at my home and their new homes.

In the early years, before a truck came to the business with pallet loads of batteries, batteries were one of my biggest headaches. I could not invest in a pallet load of batteries because I could not sell them quickly enough. The batteries came fully charged. Over a period of time the batteries self-dis-charged. First of all, my customers needed fully charged batteries to start their new solar electric systems off right. Second, discharged batteries that sit for a period of time become resistant to charge and require a maintenance

My expanded battery bank.

charge to repair the damage. My first batteries were transported by my little Dodge Colt car, 120 miles from Boston. Each battery weighed a very dense and awkward 65 pounds. Six or eight batteries had to be carefully placed throughout the car. If they had all been in the hatchback, the rear bumper would have been on the ground.

At the point in time that I needed the 24 batteries for my system expansion, I was buying batteries from a large golf-cart dealer east of Hartford, Connecticut. I was driving 90 miles to get the batteries, but the price was much better. I had replaced the Dodge Colt the year before with the Ford Bronco II, rated to carry 900 pounds of cargo. I made one trip to the distributor for the first sixteen batteries. I picked up my final eight batteries a month later while I also filled a customer's order.

The batteries were easy to install electrically. I wired them the same as the first eight batteries. Groups of four in series made large 24V batteries. I made more cables and now had six groups of four, instead of only two groups of four. What did change was the physical installation of the battery bank. The eight-battery, battery bank weighed 520 pounds. The new battery bank weighed 1,560 pounds. I made a large table-like structure of ⅝ inch plywood on 2x6s. Each end of the table was supported by walls. The center of the table was supported by two posts below.

The installation was well designed. I was proud to have it featured in books and magazines. I had no snaggle of wires that one could normally find in such pictures. Architecturally, the house and the modules looked good, both individually and collectively. There was one significant change to the workings of my house caused by my new installation: the modules partially shaded the downstairs windows like a small awning, even though they were mounted above the windows. These windows were my source of solar heat.

I first noticed the effect in the spring. The sun got higher in the sky, and the partial shading began. The effect of the shading on the solar heating effect was such that in late spring the sun did not heat the house as much as it had. I was more likely to start a short fire in the stove in the evening. There was also a beneficial result. The modules shaded the windows in the summer and the early fall. This shading prevented any direct sunlight from entering these windows in the summer. The modules greatly reduced the direct sunlight in September. This action was very important in years when we received a hot beginning to September. Because wood heat is relatively free for me, and air conditioning is not a possibility, I have found it easier to burn a little wood in April to have a cooler home during a hot August or September. I wish I could claim to have designed the shading on purpose.

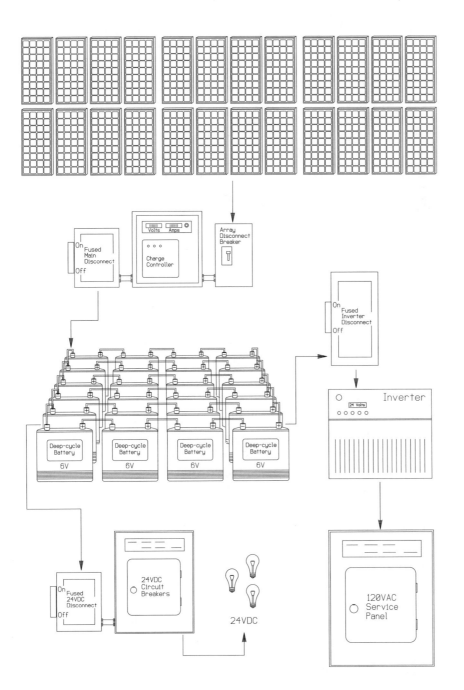

My third solar electric system.

CHAPTER 21

The Laptop Store

MY LITTLE DOG, Rolo, died in November 1986. Rolo was bouncy and energetic. Two months before he died, and two months before his fourteenth birthday, he was mistaken for a puppy by some college students. He was a perpetual puppy in stature and energy. One Saturday, Rolo had run full speed up and down the trail to the top of Bashan Hill. The next day he was sick and old. He could hardly walk. Gerry, my vet, was too attached to Rolo. Together we mistakenly passed up the logical choices that would have let Rolo die peacefully in the corner at home. Ultimately he wasted away over the next few days while Gerry and I expected a miracle.

I brought Rolo home for burial on a Sunday morning. Soon a strange car drove down my driveway. I jumped to the incorrect conclusion that this car carried another bold Sunday solar electric customer. This was the last thing I needed. Instead it was Gerry. He had been called at home by the hospital. We were two grown men too choked up to really manage to say much. I buried Rolo under the apple tree next to the garden. Today the apple tree trunk is a bird perch and bird bath. My wife Lea has turned Rolo's place into a beautiful flower garden.

Rolo had a zest for life which only ceased when all of his systems completely failed. There were some minor indications that he was ill. For the last month he had trouble holding his urine. The doggie hospital was sure that it was a hormonal problem. He had no signs of serious illness. He was an uncommonly healthy old dog. For two years he had what seemed to be a behavioral problem of occasionally throwing up after he had eaten. Gerry did an exploratory operation on Rolo. We agreed that Gerry would not bring him out of the anesthesia if the findings were bad. Gerry called to say that Rolo's entire

Jeff Fowler and Rolo, six months before the little dog died.

insides were atrophied and scarred. He was sure this had caused the vomiting problem. Gerry felt that this condition (probably cancer) had been there for years. Medically he knew Rolo should not have been alive for the last two years, much less healthy and active. Gerry could not let Rolo die that night. He felt he did not know enough about medicine to understand what was happening. Rolo had stayed alive long after his insides had failed because of his love of life, his devotion to me, and because of my love for him.

For me, life seems to have had a thread running throughout it. It is hard for my western mind to believe in reincarnation. Ideas like this are tempting, but they are just beyond my grasp. I feel that people, and emotions, and turns in life do connect. Perhaps it is just easier for me to pass through life spinning a thread that is not necessarily there. My imaginary threads are the guy wires that hold my windmill in place against the attacks of chaotic Don Quixotes.

MY MOTHER DIED much sooner than I would have expected. Mother was 60 years old. I was 24. I had always thought that the passing of my parents would be merely an inevitable event. I was quite sure that I would be too independent and too aware of the practicalities of life and death. I felt secure that I could not be shaken. These were the thoughts of someone who could not yet conceive of the reality of the future event. Like most sons, I was devastated when my mother died. There was the universal shock that the final essence of the umbilical cord was at last lost. What I felt was what most children feel when a parent dies. The only difference was that when it happened to me, it was painfully real.

I got my little dog Rolo ten months after my mother died. A young puppy, he bonded to me like a child to its mother. This bond was contagious. I began to feel a connection to my mother through him. I did not feel that Rolo was the reincarnation of my mother. Rolo appeared to have a little bit of my mother in him. Maybe in a world beyond, my mother had pointed him in the direction of my life, with a message of her love, and an essence of her personality. Of course, the real connection was love. I loved my mother, and I loved that little dog. They were similar because they both returned my love with love and loyalty.

During the late winter and the spring of 1989, I had worked long hours writing *The Solar Electric Independent Home Book*. One day, the week after my copy had gone to the printer, I sat myself down for personal direction counseling. I told myself that it was time to get out into the world. I was 41 years old. I would never have a chance of a relationship if I sat at work on Bashan Hill every day. I spent the most hours every week with my friend and assistant Ron. Ron was a bit older than I. He had found happiness later in life with a second marriage and a loving family. He always kept an eye out for some potential woman for me. He hoped I would find the same happiness he had.

I counseled myself to leave the business more with Ron. My immediate solution was to leave my house and business every Wednesday afternoon for Northampton. In this manner I would at least physically make myself available to meet someone. In the past I had desperately forced myself to visit bars. I hated bars. I was not much of a barfly. I was bored sitting on a stool trying to make my beer last as long as possible, since I did not like it much in the first place. I never went out with anyone I ever met in a bar. I did not like women who smoked or drank too much. The kind of women I would appreciate would not likely be hanging out in a bar.

My very, very cautious friend Robert, who was also 41 and unmarried, told me over and over that I only needed to meet people. I should be very careful. I should not get involved or feel any responsibility of involvement. I stripped Robert's advice of its overcautiousness and followed it. I decided that no expectations were necessary. I should allow myself to mix and let the world take its course. I enjoyed my Wednesday afternoons off. I typically made a bank deposit and ran a few business errands. I usually ate a delicious lunch in Northampton, the city of restaurants. It was springtime. I sat outside with a cup of decaf and a newspaper. I was friendly. I met women. I took no phone numbers and gave no phone numbers. Nevertheless, it all seemed to be falling into place.

I ENJOYED the Wednesday afternoons of April and May. The Wednesday afternoon of May 31, 1989 was warm, sunny, and relaxing. I dropped off my bank deposit. I ate a delicious lunch at Paul and Elizabeth's Restaurant. For an extra half an hour I sipped my bancha tea. I had gotten an early start. Freedom was there for the wandering. I chose a little excursion to the computer store ten miles away in Amherst. I had recently traded my older business computer to Ron. I had been daydreaming for months about purchasing a laptop computer to replace it. The biggest and best computer store in the area was Validata in Amherst.

I felt good all over. It was a perfect clear sunny spring day. I was relaxed with a full belly and a brain full of relaxing tea. My favorite jeans and my favorite red pocket tee shirt fit just right. It was definitely a day when Jeff's bell was ringing as I entered the laptop store. My eyes slowly adjusted from the outside bright sunlight to the inside store. The inside world melted into focus. All I could see was a young woman in sunglasses, her image so radiant that she seemed to be bathed in the outside warm sunshine while the rest of the store and people hid in their dark shadows. She seemed to be staring at me.

I smiled and sat down in a chair at the round display table of laptops. I glanced at the young woman again. She was definitely staring at me. She seemed secure that her eyes and stare were hidden behind her dark glasses. The motion of her head gave her away. The next time I looked up she was pulling out a chair to sit at the laptop table. I asked her if she was looking at laptops. She quickly and nervously replied, "No. We already have one." The word *we* instantly jolted me back to reality. She and her husband, "we," had one. I was crushed. Something had seemed so real before that statement.

I would have sworn that my disappointment had not shown, but it must have. My whole face must have dropped. I was caught completely off guard by the *we* word. My beautiful young lady became nervous at the sight of my face. She realized she had blundered. This pretty woman was not, and is still not, one to stand back and watch the dominoes fall. She innocently and adolescently blundered on. "I learned about computers many years ago when I was at Dartmouth. We have a laptop at work. I don't know much about them now. I just recently re-entered the workplace after my divorce!" She made it clear. She was not going to watch me wander off under the misconception that she was married.

We only spoke for a few minutes. Before we could introduce ourselves, or even say where we worked, we were interrupted by a laptop salesman who spoke to me, and another salesman who helped this woman out of the store with a dolly load of computer components. I dumped the computer salesman as soon as possible. I left the store to search the parking lot. She was gone. I had one clue. While we had talked, I had read, upside down, the company name and address on the invoice that the woman had in front of her at the table. She worked at HRD at a 22 Amherst Rd. I rushed to my car and maps. There was no Amherst Rd. in Amherst. I drove to nearby Amherst Rd. in Shutesbury. Number 22 was only a home. Next I went to the nearest pay phone for a phone book. There was no HRD listed. All seemed to be slipping away.

Somehow I knew this was not a fluke. It was extremely important for me to find this woman. I had one choice left. I returned to the computer store. I found my salesman. I let him jump to the conclusion that I was about to buy a couple of computers from him. After he became buddy buddy with me, the potential bringer-of-commission, I dropped the eight ball. I told him I had befriended a lady in his store only to have him interrupt our interlude. He apologized. I asked him for the name and address of the business from the invoice. He hesitated. Finally he provided the information: Human Resource Development Press, 22 Amherst Rd., Pelham, Massachusetts. I thanked him and immediately pressed him for her name. My salesman told me he did not know her. I suggested that she must have signed to receive the goods she had purchased. I had him. He returned with the name Beatrice A. Sikora.

On the way home I purchased a pretty card. The next day I related the whole story to Ron. He was excited. He told me I had to send a card. The woman would not have said what she had said unless she had been very interested. In the card I wrote, "I enjoyed meeting you at the computer store on Wednesday. I am sorry we were interrupted. Would you like to get together for ice cream, or perhaps a drink, or perhaps a glass of carrot juice?" I posted the letter that Thursday. Early Friday morning I drove to Boston to give a Saturday morning lecture at Northeastern University. Saturday evening, I returned home and read the business telephone log. Ron had recorded a personal message from my personal telephone line. The entry was marked with big red stars all around so that I could not miss it. Leatrice (not Beatrice) Sikora had called at 11:10 Friday morning. She had obviously called immediately after she received my card in the morning mail at work.

I phoned Leatrice immediately. The background on her end of the conversation sounded like the Sunshine Day Care Center. Leatrice was interrupted three times by the children. We spoke for an hour in three different sessions over the next two hours. Leatrice was happy I had called, but also nervous. Her first insecurity was that I would hang up upon hearing that she had three children. She was so nervous explaining that she had three kids, and that her husband had physical custody, that it sounded like the kids had been taken away from her because of poor care or abuse. She told me she had accepted it and that things were going well now. I heard my friend Robert's warnings of caution. However, Leatrice, who called herself Lea (pronounced *Lee*), and I were not cautious plodders like Robert. We chose the first available time, Monday evening, to meet for dinner in Northampton.

Lea was nervous at the restaurant. I had picked Chinese food, with which she was not very familiar, and I had ordered chopsticks, which she had never used. Soon she relaxed and we had a wonderful dinner that turned into a long walk with ice cream to Paradise Pond on the Smith College Campus. Once relaxed, Lea told me the clear story of her recent life. She had been divorced for two years. During the bitter divorce her rich husband had used all his power and resources to gain day-to-day custody of the children. Ironically, during their whole marriage, he had been the busy doctor who had hardly seen the kids. The kids' whole world had revolved around Mom. Lea felt very guilty that she had not been able keep her children. She was terrified that people would assume that the children were taken away for bad reasons. It was this fear that had created my misinterpretation of our first conversation. Lea and I were fast friends and confidantes. We made a date for next Saturday night.

The following Saturday was a workday for me. Bruce Wilson, the sales manager from Atlantic Solar Products Inc., had an appointment to visit my home and business. I expected him to arrive in the late morning and leave by early afternoon. I had invited Lea to have dinner at my house in the country. I waited, and I waited. No Bruce and no call from Bruce. Waiting always makes me nervous. I began to get nervous about Lea coming to dinner. I wanted to see Lea; I was nervous about her coming to my house and me having to be the guy who built the passive solar house and had the solar electric business. I wanted to be just me. I called and asked if I could go to her house.

Bruce did not show. At five o'clock I left a note on the door and left for Lea's house in Greenfield. As I drove up Bashan Hill Road, I was confronted by a speeding car containing Bruce and his friend. They had obviously been having a fine party day. I held my emotions and invited them back to the house. However, I told them I had plans that could not be changed. I dumped Bruce in half an hour and headed for my new friend Lea.

Poor Lea had gotten so many phone calls that day from me changing plans, then making them later and later, that she was sure I did not really want to see her. She expected we would go out for dinner. I shocked her when I suggested a trip to the market for food and wine, and dinner at her house. Lea thought she was about to suffer a fate worse than chopsticks. She would have to cook for this new friend. Of course, I planned to cook all along. I love to cook. From that day on Lea and I have been together. The next day we drove to Worthington. She loved it. My part of Bashan Hill

Leatrice Sikora, soon to be Leatrice S. Fowler.

looked like the Vermont where she had grown up. The thread of my life was meshing with the thread of her life. Soon she would be my wife.

Lea worked for a publisher in Amherst, a long drive from Greenfield and a very long drive from Worthington. Lea was a world class long distance runner who had been injured the previous fall. She could not seem to heal. For several months in the fall, we drove to Boston for her physical therapy. Unfortunately, the ride to Boston and back was as bad on her hip injury as the physical therapy was good. The ride to Amherst was also taking its toll. We decided to accept an offer to sell Fowler Solar Electric Inc. to a large east-coast solar electric distributor. The deal called for me to develop a larger version of Fowler Solar Electric Inc. for them. I would work from my office. Lea could be my assistant and retire from the long drives to Amherst.

The business sale was to be completed before Christmas. The distributor then pushed the date up to the beginning of the year. Finally, one day in the middle of January, we received a copy of the contract from the distributor's lawyer. The written agreement was bad, and totally different from our hand-shake agreement. My lawyer called. He had received his copy earlier in the day. We both felt that we could not trust the people who had sent this contract. The contract was so contrary to our verbal agreement that even if I negotiated out the bogus elements, I would still be in business with the people who had tried to pull the fast one. My lawyer had advised me earlier that no contract for the sale of my business would be easy to enforce. The basic element of security of the sale was that I would be able to trust the people with whom I would be working. The trust was gone.

Lea and I left the business to the answering machine and went for a long walk. By the end of an hour we had made our decision. We would not sell. Lea would work nearly full time for the business. This would be the right kind of additional labor that I needed. Lea and I had been working for three months on the plan to create an expanded Fowler Solar Electric Inc. for the large distributor. We had kept our cards close to our chests. After the walk we wrote a short letter to the potential buyers. We told them that the contract was unacceptably different from what we had agreed upon verbally. We needed to get our yearly catalog out on time, and we did not have time for further negotiations. The deal was off. We would continue to buy modules from them. Business as usual.

The next day Lea and I started our work on the expanded catalog and the marketing campaign that we had designed for the large distributor. We were happy as clams. We were excited about our joint venture. Lea and I spent all of our days together. We worked together at the business, and we ran together at the end of the day. On the weekends we worked together finishing the house. We learned that we had to deal with the petty differences of our relationship differently from how a normal couple did. We could not have a tiff and then storm off to work to get away and later return with all forgotten.

We needed to settle our differences quickly, kiss and make up, and go on with our day.

When I first met Lea she was not doing well with her ex-husband. His anger often erupted in his driveway when Lea was picking up the kids. It was difficult to schedule the children's custody visits. When Lea worked full time at HRD Press, she had little time other than weekends that was compatible with the children's schedule. Together we worked to better the situation. Lea's job at Fowler Solar Electric Inc., by design, worked around the custody schedules of the kids. She could now take the kids for weeks in the summer and on school vacations. Whenever possible we both drove to Greenfield to pick up and return the kids to promote a peaceful transition. Life was good on Bashan Hill. Solar electricity provided our home with power and provided Lea and me with jobs that provided our children with a second home. All provided me with a family. Someday I may even get around to buying myself a laptop.

Step-daddy and Jarod, Kurt, and Bethy.

CHAPTER 22

The Garage, or the Great Wall

THERE ARE SEVERAL pieces of land on Bashan Hill road that can be considered building lots. Each time one of these lots has been put up for sale, a real estate agent came with a sign and hammered it into the ground at the side of the road. When the sign was close to my house, I investigated the price. I bought one seven-acre piece of land across the road. The other parcels have always been too expensive. The lots immediately across the road from the end of my driveway were not attractive. Each lot had 20 to 40 acres but only a few hundred feet of road frontage. Each lot was extremely long and narrow, and only the first few acres of each lot were accessible. The remainder of each lot was a narrow but impassable section of wetlands backed by a long, narrow, inaccessible woodlot. The real estate agents priced the lot by total acreage, not usable acreage.

Most people did not want to pay for a full 20 acres of land to only be able to use two acres of it. If they were willing to buy the whole deal, they would not have the money left over to bring in a power line. Without the power line, the potential buyer could not secure a mortgage from a bank to buy the land and build the house. The land along the first mile of Bashan Hill Road is not subdivided into lots. There will be no economic pressure to do so until Worthington is such an attractive place to live that people will build on ledge and the town will provide a municipal sewage system. Because of the inhospitality of the first mile, the power line had not been extended by increments as it had been in the rest of town.

As long as the power line does not extend down Bashan Hill

Road, it will be unlikely that I would have a new neighbor. No other parcels of land on this road were suitable for a homestead, other than the Cooks' land and my land. I considered trying to buy the lots at the end of my driveway. For most of the time that I have lived here, I did not have the credit to borrow money from the bank. Now I have a child who will need an education.

Most of the real estate agents who tried to sell land on my road were not familiar with Worthington. They priced the land too high, and they were too lazy to show the land to potential buyers. Consequently, the buyers always seemed to show up at my door or the Cooks' door. Most were shocked to learn that there is no power line. We quoted them the price of a power line extension. We told them we were not willing to split the cost of an extension. We matter-of-factly explained life with alternative energy. We did not bother to explain that we liked it the way it was on our road. Soon the realty sign would be buried by the snowplow and lost in the growth of the spring ferns.

In the summer of 1986, there was a 20 acre piece of land for sale on Bashan Hill Road across the road from Bob and Karin Cook. It was one third of a mile up the road from me and not my concern. The previous year, there had been a 26 acre piece of land for sale across the road from my driveway. The few potential buyers who had looked at it had never returned. The realty sign was gone. I assumed it was safely off the market, until a man and woman arrived at my house. The man introduced himself as Jeffrey Hartmann, not to be confused with Jeffrey Hartman who was my friend from Boston. These two people were much different.

Jeffrey Hartmann introduced himself as if he were a friend of my neighbor, Bob Cook. He said, "Bob Cook sent me to talk to you about solar electricity for the 26 acre piece of land across the road from you." I erroneously assumed he was a good friend of Bob Cook. Bob had told me during the last winter how he had spoken to a good friend of his who had considered building an earth-bermed house and homestead on Bashan Hill Road. All of my defenses were down. I should have been a tight-lipped Yankee. Jeffrey Hartmann was not a friend of Bob's. He had met Bob the day before while looking at the other 20 acre parcel of land for sale.

Against all logic, Hartmann bought the land. I was always aware that there was a possibility of another home on Bashan Hill Road. I was well aware that I had invaded the Cooks' domain several years before when I became the second home on the road. I had hoped that the eventual neighbor would be someone with a lifestyle similar to mine. I was not really prepared for the first new neighbor on this long and rural road to land at the end of my driveway. I hoped for the best. I tried to become friends with my new neighbors. He was a nervous type, but his wife was very kind. I coaxed him into siting his potential house two hundred feet off into the trees, where he could have his privacy and I could have mine.

The possibility of
neighbors.

Only the road frontage and first half acre of land that Hartmann had bought were in Worthington. The remainder of the land was located in Cummington. I was part of the Worthington town government. Soon I heard complaints from both Worthington and Cummington officials about my neighbor, who cried "poor me" about every potential license and permit. Hartmann complained to me that he was being picked on by the towns. Hartmann brought in a ratty old mobile home. Cummington found out about it after the fact (not from me). Town officials chastised him but instantly handed him a permit without an inspection. This set the precedent for the future. Hartmann rarely took out the appropriate permits. The Town of Cummington issued him permits retroactively. As time passed, the junk piled up, and the number of dead vehicles along the narrow road grew. I remained on good terms with the Hartmanns and I tried to help them with advice when they asked for it. Life in the hills was hard for them. They did not have the skills to build a homestead. Ultimately it was the New England winters on Bashan Hill that did them in. The nice wife and daughter refused to live here anymore.

Once Jeffrey Hartmann had to apply for the trailer permit, he also had to meet the basic requirements of the local Board of Health. These were a septic system and running water. To have running water, he needed a solar electric system to power a well pump. I sold Jeffrey Hartmann a ten-module solar electric system. In my own mind, I realized that a neighbor was here to stay. A trailer could go, or collapse to a junk pile from winter snow load, but the well and septic system were here to stay. They added value to the lot. If Jeffrey Hartmann left, someone else would follow. The next owner could be worse.

I had always considered building a garage. I owned a tractor that was garaged under a blue tarp. I repaired my own car, sometimes in a snowy driveway. A garage was a logical addition to my homestead. I had not built a garage because I had neither the time nor the money. When it became apparent that I would have a neighbor, I began to seriously consider one. My house was placed 200 feet off the road amongst 25 foot tall balsam firs. These trees provided me with complete visual privacy from all angles, except the line of vision from the road to my house, down my straight driveway. If I had been more clever, I would have curved or snaked my driveway. Jeffrey Hartmann had an ungraded driveway. He found it easier to drive farther down the road in his truck and turn around in my driveway. A minimum of twice a day someone was driving into my driveway and looking down upon half of my homestead. This was not the rural privacy to which I had been accustomed.

The day Jeffrey Hartmann ordered his solar electric system, I made my final decision to build my garage. I planned to build it at the house end of the driveway. This garage would be my New England version of the Great Wall of China. The garage would block the line of vision to my gardens, ter-

The view down my drive-
way is now blocked by a
garage.

race, and homestead. Later, if more houses were built on my road, I could
relocate my driveway, snaking it through the trees, and transplant trees in the
current driveway. I felt I had an adequate plan for rural privacy that circum-
vented the need to purchase all the lots across the road.

I realized nearly $2,000 of gross profit from selling solar electric compo-
nents to Jeffrey Hartmann. I earmarked this sum for my garage project. I
hoped to allot $2,000 more out of my future business income. I would be
able to build the garage for $4,000 if I purchased local green framing lumber

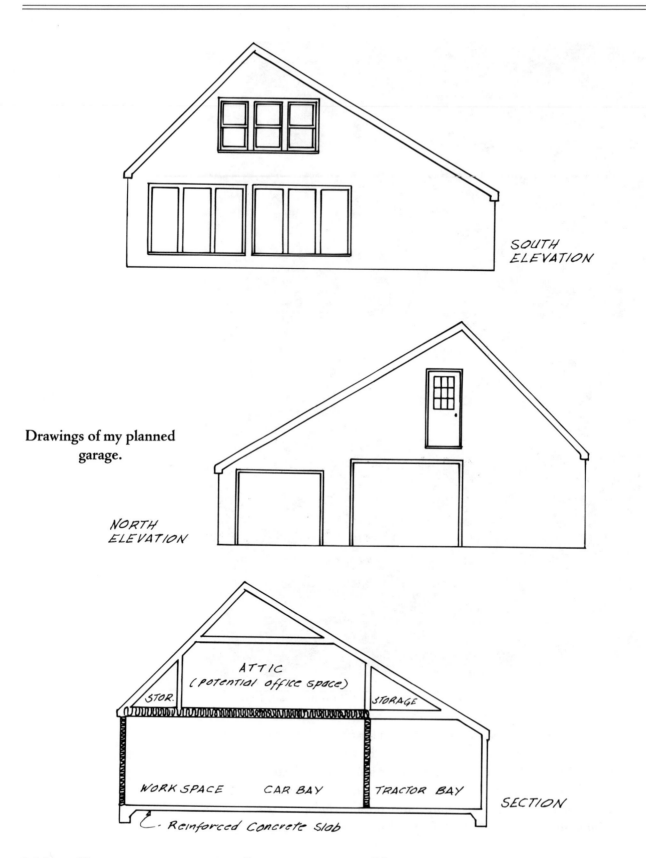

SOUTH
ELEVATION

Drawings of my planned garage.

NORTH
ELEVATION

ATTIC
(potential office space)

STOR.   STORAGE

WORK SPACE   CAR BAY   TRACTOR BAY

SECTION

Reinforced Concrete Slab

and native pine siding. I would build the garage on weekends. This project was new and exciting to me. I had built the house with a generator but I would be able to construct this building from the power of the sun.

When I committed myself to building the garage, my first thoughts were to build a simple and inexpensive building that could be closed in before winter. This was a foolish idea. The building plan would be rushed, and I would be overcommitted at the busiest time of the year, when my New England customers rushed to install their new systems before winter. I decided to take the entire winter to develop my plans. The longer I thought about it, bigger seemed better. A larger garage would be a more effective visual wall for the site. My solar electric business was steadily growing. I could see the day when I would want the business offices out of my home.

A COLD UNHEATED GARAGE in the winter in New England is not much more pleasant to work in than a snowy driveway. At least in a snowy driveway, you might be able to pick a sunny day. Most people in Worthington heat their uninsulated garages with wood stoves. This requires a lot of wood and a lot of planning. The garage starts out at the outside temperature. The wood stove must run a long time to raise the temperature in the unheated building. At night the temperature drops. The next morning, the whole process begins again. My first decision was to build an insulated garage with an upstairs attic space that could be insulated and finished in the future. If at any point the business should need to move from the house to the garage, the upstairs could become the offices, and the downstairs could become the storeroom.

One of my earlier dreams had been to build an earth-bermed, or underground, garage. An underground building will maintain a temperature of 50°F. in the winter with no heat. I had liked the idea for my homestead. The garage would have been cool in the summer and warm enough for garage work in the winter. The underground garage would not work at my site, though; I had no side hill in which to tunnel, nor a landscape that would accept an aboveground earth-bermed building. I still liked the idea of a garage that needed no heat.

I considered a passive solar garage. I could make a wall of windows on the south wall. Lumberyards stocked a standard size insulated glass panel for replacement in sliding glass doors. Economy of scale priced these low. If I framed these insulated glass panels into the south wall, the wall would not be much more expensive than a conventional wall of studs, insulation, and siding. The walls and the ceiling of the garage would be insulated. The floor of the garage had to be concrete and would serve as the thermal mass for heat storage. I had monitored my home for the past several years. When I was away during the winter, the house always stayed above 45° F. A passive solar garage could never be as warm as my home without the added expense of a fancier foundation and underground insulation for the concrete floor. My heat storage could only be fair.

My final plan was to build a passive solar garage that could maintain a temperature around 30° F. at night and on cloudy days. This would be a basic tolerable temperature for carpentry work or car repair. I would hope to save my garage work for sunny days when the winter sun would raise the temperature to a range 50° F. to 70° F. If my passive solar heat estimates were low, I could always add a wood stove to be used during the coldest times in winter. This insulated building would be easy to heat.

Conventional overhead garage doors fit poorly. Even the insulated models lose too much heat. I planned to use hinged dual doors. I would custom build these doors with adequate foam core insulation. The building would have adequate interior space for the doors to open inward, thus avoiding the problem of opening outward against deep winter snows. In the house, I meticulously maximized the insulation to provide comfort in my living space. I chose to use only basic insulation for the garage, and to insulate only the downstairs. I planned to use the garage a minimum number of days in the winter. For now, upstairs would remain an unheated and uninsulated space.

I wanted to build this garage once. It is not easy to put an addition onto a building that is built on a slab. Originally the building was to be a standard 24'x24' two-bay garage with a 45° roof pitch to give adequate space for an office above. As the planning progressed, I added another twelve feet to the garage for a tractor bay. This area would be uninsulated, unheated, and have no south facing windows. The change in the roof angle that resulted from the addition of the tractor bay added extra space for the potential office upstairs. I now had plans to build a 24'x36' building with a 20'x24' attic or potential office space. The 36 foot dimension of the building would run perpendicular to the end of the driveway and create a 36 foot wall.

Early in the spring, to make room for the garage, I regretfully cut down the tall cherry tree that I had climbed the first day I had visited the land on Bashan Hill. The area where the garage was to be built was already scraped free of topsoil from the original house construction. It had been used as a parking area. The subsoil was low grade gravel. This would provide adequate drainage. I ordered 36 yards of bank-run gravel. I graded and compacted this gravel with my tractor and bucket loader to form a one-foot-thick level base for the garage slab. Most garages in our area are built on a foundation that goes below the frost depth of four feet. The frost does not get below the foundation and cannot lift the concrete wall and crack it. Inside the foundation, a concrete floor is then poured.

I chose to build this garage on a slab instead of on a foundation. Typical slab construction consists of a one-foot-thick slab of concrete on the ground. The construction is cheap in labor, but expensive in terms of concrete. From the recommendations of articles and Bob Cook, I chose to build a six-inch-

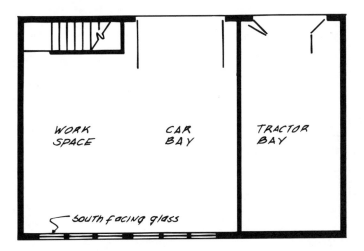

GROUND FLOOR PLAN

WORK SPACE

CAR BAY

TRACTOR BAY

south facing glass

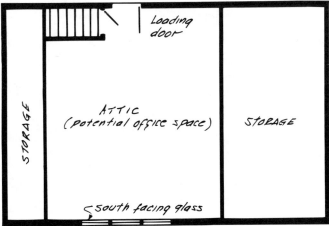

SECOND FLOOR PLAN

STORAGE

ATTIC
(potential office space)

STORAGE

Loading door

south facing glass

Floor plans for the garage.

thick reinforced concrete slab. The compacted and molded gravel was the bottom of the form for my slab. Perimeter planks were the edges of the form. The form around the perimeter of the slab was deeper, forming a narrow ditch around the inside of the perimeter of the form. The concrete poured into this area would later harden into an 18"x8" inch beam around the whole perimeter of the 24'x36' slab. This would act as the foundation to the walls of the garage.

I had learned about concrete reinforcing in my house project and designed the garage slab for maximum strength. This slab would rival the strength of a bridge. There would be no frost wall and foundation beneath this slab. The ground would freeze under this slab and lift it up. With no reinforcing rods the slab would crack. I constructed this slab such that a large crane could

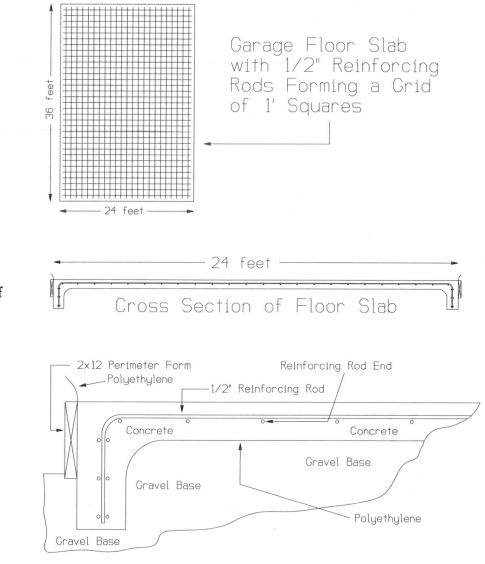

Garage Floor Slab
with 1/2" Reinforcing
Rods Forming a Grid
of 1' Squares

36 feet

24 feet

24 feet

Cross Section of Floor Slab

Construction details of the garage reinforced concrete slab.

2x12 Perimeter Form
Polyethylene

Reinforcing Rod End

1/2" Reinforcing Rod

Concrete

Concrete

Gravel Base

Gravel Base

Polyethylene

Gravel Base

Magnified Cross Section of Floor Slab

pick up one corner, and all that would happen would be that the building would tip. The building was designed to be lifted by the frost in winter and to settle back in place when spring came, resulting in no harm to the slab or the building.

I molded the gravel base of the form such that it would create an eight-inch-wide by eighteen-inch-deep perimeter grade beam that would be continuous to the interior six-inch thick garage floor. On the floor of the form I placed a six-mil layer of polyethylene to contain the wetness of the concrete until it hardened, and to provide a barrier from the wicking of ground moisture up through the garage floor. Bob Cook came down on a Sunday with his transit to level the forms and help tie the reinforcing rods. We tied half-inch reinforcing bars, one-foot on center, in a giant grid to reinforce the floor of the garage. This was much more heavy duty than the standard sheep fencing that is normally used. We placed six, half-inch rods all around the perimeter of the building, inside the grade beam portion of the slab.

The following day we poured the concrete. This was a job that I could not do myself. Whenever one of my friends is about to pour concrete, there is a call to arms. I issued my alarm the previous day. Bob Cook, Dave Whitcomb, and Jonathan Ginzberg showed up exactly one-half hour before the scheduled arrival of the concrete truck. I passed out shovels and rubber boots. For the next hour, all of us worked at a feverish pace shoveling and leveling the heavy concrete. We sat down in a sweaty exhaustion to drink juice and water. Jonathan and I waited for the concrete to set up. We then troweled the surface with a gasoline engine–driven power trowel to create a smooth floor finish.

Jimmy Molyneux — friend, real estate agent, and building inspector — outside of the Corners Grocery.

I had no time during the week to work on the garage. I hired my nephew, David Beach, for most of the Saturdays that summer. David was 20 years old, but an experienced carpenter. Scott also worked a few Saturdays. I worked on Sundays by myself. Clarence Witter had stopped selling lumber because of a chronically bad back, so I bought my framing lumber from Roger Gunn, another Worthington resident. The building progressed uneventfully. The garage steadily grew in size each weekend all summer long. By fall the building was framed, sheathed, and roofed. On Saturdays during the fall, I installed the windows and made the doors. The building had no siding, but it was closed in for the winter.

The garage is 20 feet from the house. The battery bank in the basement was 40 feet from the garage. This meant that any low-voltage DC circuit to the garage would have had to run 40 feet before it could even start its long run in the garage building. This would have meant large and expensive custom wiring if the battery bank in the house had been used as a source for DC lights in the garage. I did not want a separate solar electric system in the garage. I had enough electricity from the house system to power the garage. The only logical choice was to wire the garage for only 120VAC.

Naturally, Bill Wilson laid out the wiring in the garage. We planned for the future expanded use of the building, if someday the business moved to the garage. He, of course, let me wire almost everything. The largest loads would be power tools. Light would be provided by efficient, compact fluorescent bulbs. The wall of windows in this passive solar garage would bathe

The south side of the garage.

the work area in natural light during any day. This had been one of the convincing arguments for the passive solar garage in the design stage. Most garages are dark. To work on a carpentry project or on a car in a dark garage, you need a lot of light. It is inconvenient to be moving the light over your work or your work to the source of light. I would have natural light everywhere all day long. I could save my electricity for other loads.

MOST OF MY NON-LOCAL FRIENDS who live in Worthington moved here because they loved the community. There was no source of employment, so they carved out their own little niche. My friend Jimmy Molyneux was typical. He and his wife Penny managed to afford land through a group purchase of a large tract of land. They built their own home up around them as they could afford the materials. Penny started a home-based bakery. Jimmy worked as a contractor and painter until he became a licensed real estate agent under the wing of Mary Jane Carey, who sold me my land. The fall of my garage project, Jimmy was also the town building inspector.

One day soon after my garage was framed and sheathed, I ran into Jimmy at the general store on my daily run to the post office. Jimmy always kept me informed of any potential real estate news on my road. The previous spring, he had alerted me that he had three lots for sale across the road and down the road from my land. The nearest lot was diagonally across the road from the far end of my land. This seven-acre lot was not a threat to my privacy, but still I negotiated to purchase it. The land abutted the camp of Phil and Ann. At the eleventh hour I let them purchase the land to increase their lot size and gain access to their back acreage. Earlier in the month, Jimmy had told me he thought he had sold the second seven-acre lot to a man who wanted to grow balsam Christmas trees. This fall day, I got the news that there was no firm commitment on the land sale.

This seven-acre lot was for sale for $10,000. It was not yet perc tested for building. I asked Jimmy, "What would happen if I wanted to buy the land right now?" Jimmy replied, "It is for sale with no obligations to the other buyer." I returned home to drop the mail at the office and went for a walk to the land. I called Jimmy and told him I would buy the land for $8,000 and pay for any perc testing I required. A perc test cost $400. I planned to not build on the land anyway. I drove my tractor into the woods, dug a hole, and performed my private backwoods perc test. The land had an excellent perc rate, as I had expected from the lay of the land and the flora and fauna.

I finally took ownership of my new seven acres late in the winter. I had actually purchased the land for three reasons. This would be one fewer lot on the road to become a house lot, I needed a woodlot for cordwood, and this land had at least 10,000 board feet of pine that was easily accessible. The owner had never seen this land. He had had the land logged before he broke it up into smaller lots. The loggers had only logged for hardwoods. They had never touched my lot because it was mostly pine. As soon as I assumed ownership of this land, I cut 2,500 board feet of pine logs and hauled them to Bob Cook's small sawmill. I stickered the sawed pine. Later I used the dry pine boards for siding on my garage.

Finishing the garage turned out to be more fun than I had expected. While the stickered pine boards were air drying, I found my true love and carpenter's helper at the Amherst computer store. Lea and I shiplapped the pine boards for the siding on the garage. Both edges of each board had to be passed through the dado blade on the radial arm saw. I ran the center of the board and the saw while Lea guided the other end. We worked at shiplapping in half-hour doses. Our arms got tired, and the inverter heated up from the long and heavy load. We quit before the inverter shut off from overheating, and before we got tired and sloppy while working with a dangerous power tool. In between the half hours of shiplapping, I installed the siding boards while Lea followed with the screw gun to screw in the galvanized screws.

We left the garage at this stage until it was later necessary to move our offices from the house to the garage. We did not even bother to insulate. The garage was already too full of solar electric components, lumber, and a tractor. The garage stayed adequately above the outside temperature from the solar gain of the windows even before we insulated. I loved my new garage. I had a very large building for less than $5,000. The whole project had been built with solar electricity. The majority of the materials were from my land and other land in Worthington. The building of this building had been more environmentally sound than the building of the house. I had a beautiful carpenter's helper. Life was good on Bashan Hill.

CHAPTER 23

Solar Electricity as a Profession

T HE NEATEST summer jobs in Worthington, when I was a teenager, were haying or working at the potato farms. A kid could work all summer without a shirt in the hot sun. He could drive tractors and farm trucks around town long before he was old enough to have a license. He could even drive to the general store on a tractor for sodas, walk in shirtless and tanned, flex his muscles as he dug his money from his jeans pocket, and sneak a look to see if the girls were watching. I worked at the golf course the summers when I was twelve and thirteen. The latter summer I worked watering the greens, because I had a tractor to drive. I had to be fully and neatly clothed. This was not as cool as working on the farms.

The best summer and after-school job in town was the job at the general store. Pete and Shirley Packard owned the store. They were only about thirty years old. They had recently bought the store from Pete's father, Merwin Packard, who had owned it since 1924. Merwin sold the store when he became the full-time postmaster. Danny Dunlevy worked at the store until the end of the summer when I was thirteen. Danny was the older brother of my friend Paul Dunlevy, and a top athlete and scholar at the local high school. He was much more than a mere bag boy. He was Pete's assistant manager. He had to be smart and good with people. He was respected by all of the adults and all of the kids, too. I have to admit that I daydreamed about working Danny's job at the store.

In September of 1961, Danny packed his belongings and headed for the University of Wyoming. He had a scholarship at a school

that was as far from home as he could find. Danny's job went to my best friend Douggie Torrey, who was only one year older than I. He was fourteen, the age at which one could work under the child labor laws. It seemed that I was a year too young at the wrong time. Douggie had been hired for the job at the store, and would of course keep it the next four years, until he would leave for college. In the fall of 1961, Pete Packard asked me to work two Saturdays because Douggie had to attend two family weddings. I liked the job. I was too young to realize that people managed to keep jobs because they were good at them. It never occurred to me that I was more suited for the job than Douggie. After all, he had always been one step above me at our jobs at the golf course. Worthington was a small town, and the Torreys were friends and neighbors of the Packards. When Thanksgiving came, Pete did not have enough work for Douggie. When New Years came, Pete hired me.

I worked in the store for eight years. The Packards were my second family. I worked Saturdays and after school during the school year and full time in the summers. I spent more time at the store than I did at home. Pete always worked my schedule around my varsity sports. Pete loved merchandising and public relations. We were always trying new things in this little country store. Pete taught me any part of the business in which I showed interest. Somehow, Pete never saw me as a bag boy or a little kid. From the beginning, I was his assistant. I remember going to sales meetings at the big distributors when I was fifteen. Pete introduced me to all as his assistant manager, and I really was. After basketball season, my senior year in high school, Pete was appointed postmaster. We shared the full-time meat-cutting job. Pete worked early mornings, and I worked 20 hours per week during the spring until I could take over the full-time meat-cutter job for the summer. During August, I trained the new twenty-six-year-old meat cutter. I managed the whole business after my junior year of college. Working at the Corners Grocery was a way of life for me. It was the best job I ever had. I was an excellent high school student, so I went to college. I am glad I have been educated and that I have experienced a world beyond Worthington. Still, I feel a loss at having given up the simplicity and innocence I might have kept as a small town boy who had never left.

Pete Packard in 1966.

I NEVER LOOKED for the big professional job after college that is the core of the American dream. I worked jobs that were interesting to me. I normally obtained an entry level job in a very small business. Soon I was the manager or assistant manager directly under the owners. Eventually, they would offer me a place in the business or a part ownership of a business expansion in a new location. I always declined their offers. After the project of building my house, I was looking for a job. I had always been very good at many things, but I wished to be excellent at one thing. It was a bit of a timely midlife doldrums. I enrolled at Antioch New England Graduate School to become an environmentalist with a marketable degree. Toward the end of my Masters program all the lemons came up in a row, and the silver dollars started jingling from the slot machine. The first lemon was my self-education in alternative energy, the second was my renewed academic skills from graduate school, and the third was my experience in small business.

My profession in solar electricity seemed to naturally evolve. I had educated myself for my own needs as a homeowner in the backwoods. After my first year living with a small solar electric system, I was certain that I would expand it. Most advertisements for solar electric components finished with a small line, "Dealer inquiries welcome." I already knew how to set up and keep records for a small business. I considered becoming a retailer of solar electric components. I called the manufacturers and distributors of the appropriate goods. Soon I realized that it was relatively easy to become a dealer in this very young industry. I knew how to set up a small business with a name, tax number, and a checking account.

Most distributors required a minimum order structured such that an individual could not easily become a dealer to obtain wholesale pricing for his or her own private use. I did not have the money to place minimum orders for all the goods that I needed to carry in order to become a real alternative energy business. Once I found a way to place the minimum orders required to obtain dealer status, I would then be able to order components at smaller quantities immediately after customers ordered them from me. My business would require no large investment because I would carry almost no inventory.

My first exposure to the world of solar electricity had come in the form of the cooperative purchase of modules organized by Joel Davidson. My first telephone call to obtain a dealership fittingly was to Joel Davidson, who became the sales manager at a distributor of solar electric modules. Joel's employer, Wm. Lamb Corporation, required a minimum order of eight modules. My solution was to draw on my experience obtained from my earlier cooperative purchase from Joel. I organized my own secret cooperative buy. I combined my personal order with an order from Bob and an order from Richard to reach the magic number of eight. I later added an order of four more modules for stock in my new business.

Heart Interface required a minimum purchase of two inverters to establish dealer status. No individual would want to purchase two inverters for his or her home. Heart Inverters wanted desperately to sell inverters through a New England dealer. I ordered one inverter for my own home, which would serve as a working display model, and one inverter for Richard. All future orders for inverters could be ordered after I had a down payment from a customer. I had already purchased batteries at a wholesaler in Boston. Smaller items, such as charge controllers, could be purchased from Joel. I could purchase electrical supplies through Bill Wilson. I was theoretically in the solar electric business.

MY FIRST CUSTOMERS were people I already knew in the hilltowns who were living without electricity. Later, customers were friends of friends. Often someone who had read an article in the local newspaper featuring my house would just show up at my door. I could afford to sell equipment at similar prices advertised in the mail-order catalogs from California. I sold all the same equipment. People bought equipment from me because they could come to my home and actually see the components in operation. For the first few years, it was a part-time business. I had no way of finding enough customers to expand the business to a level that could pay me a full-time income. I was always afraid that a larger business in the area would instantly end my small venture. I chose to earn a Masters Degree in Environmental Studies.

My choice to go to graduate school ironically resulted in the formation of a full time solar electric business. During the summer of my Masters program, I chose to do the independent study to write my how-to book. I also had a half-time job on the Wood Industry Project at our local Hilltown

Community Development Corporation. Together, these two aspects of my life set the stage for the future. I managed to write and print the book largely because my friend and boss was an early personal computer user. Derrick Mason purchased a computer for the Hilltown CDC. I worked nights and weekends on the book using the computer. I could not type, but the computer allowed me to type out my book anyway, because of its correction capabilities.

I wrote my first book, *Solar Electricity for the Remote Site Home*, in six weeks on yellow pads. I spent a similar amount of time typing it into the computer. It was a painful experience. In the beginning, I did not even know where the letters were on the keyboard. I would swear in frustration because I was so tired that I could not find the next key to be typed. By the end, I had learned to type with two fingers and to run the word processor on the computer. I learned another valuable lesson at the Hilltown CDC: I hated the state bureaucracy. I will exaggerate and say I felt we spent half the time validating to the state what we were doing with our grant money, and the other half of the time writing the next grant. We didn't spend enough time actually accomplishing anything.

I printed out a copy of my book at the Hilltown CDC. I pasted up diagrams in the spaces I had left in the text. This was long before desktop publishing became popular on PCs. The finished book contained about one hundred 8.5"x11" pages. There was no place to send it for consideration for publishing. The book was too specialized and written too independently. I found a local cooperatively owned photocopy business that could inexpensively copy text-only pages on an outdated machine. If I purchased 100 copies at a time, I paid two dollars per copy. The book copy consisted of 100 pages with a heavyweight paper cover and back page. At home I stapled each book with a giant manual stapler.

At this time I was purchasing my modules from Paul McCluskey of Solar Electric Specialties Co. in California. I sent Paul a copy of my new book. I wanted his company to sell it. Solar Electric Specialties Co. never purchased any books, but Paul provided some help on his own. He handed out copies of my book to some of his large dealers. A few weeks later, I received a call from Windy Dankoff from New Mexico. Windy had been in the alternative energy business long before solar electricity, as his first name indicates. I had read articles by him, and references to his work, in wind power books. I was shocked and nervous on the phone. "This is the book I always wanted to write, but I never found the time. I want to buy 12 copies," Windy said. Soon I got similar calls from Steve Wiley of Backwoods Solar Electric Systems and John Schaeffer of Real Goods Trading Corporation. My simple homegrown book was selling. I was no longer a backwoods solar electric homeowner, but a spokesperson for solar electricity.

The Hilltown CDC moved to its new offices in the summer while I was writing my book. The new office space was in a building converted from an automobile service garage. I was typing on the computer in the same space where Tommy Hinton had painted my 1960 Volvo in 1967. The office had all the pollution liabilities of new office space. The concrete floor of the garage was covered with synthetic carpet glued to the old floor, the walls were painted with new paint, there was no ventilation system, and we had a large photocopy machine. The building was just bearable in the summer with all the windows open. As fall came, we started to shut the windows. The office air became more disgusting. The offices smelled like a synthetic carpet warehouse.

I began to question the direction of my life. I had completed all the requirements for my Masters degree. It was time to apply for full-time employment in my new field. The environmental jobs for which I might apply were either part of state agencies or funded largely by state agencies. I could not bear to work again in such a bureaucratic climate. I wanted to change the world. Slowly, I began to realize that I might environmentally improve the world one baby step at a time with each new solar electric system that I installed and each book that I sold. In mid-October, the same day that the windows of the offices closed, I spoke with Derrick and gave my two-week notice. I was committed. I was now in the solar electric business full time. I had to make it go or sink into debt.

Fowler Solar Electric Inc. grew each month. Over the following three-and-a-half years I sold 3,000 copies of *Solar Electricity for the Remote Site Home*. I sold these books to solar electric users, who tended to lend them to their neighbors. I estimated that I had aided at least 6,000 homes. In August of 1988, the business seemed to be growing beyond my ability to run it by myself. I had shopped out all non-essential tasks such as bookkeeping and copying. My nephew made battery cables, and his wife stapled catalogs for mailings. I needed to be able to take a day off or a vacation. I needed an employee who could perform at a level of expertise similar to mine.

The post office in Worthington has been a part of the Corners Grocery for at least the last seventy years. Today, it has its own space in the Corners Grocery building which is rented by the U.S. Postal Service. The space is hardly separate from the general store. When the post office closes, the old glass window to the public is lowered and locked. The rest of the post office remains open to the public whenever the store is open. Anyone with a postal box can collect his or her mail on Sundays. My friend Pete Packard was the Postmaster until he retired in 1991.

The Corners Grocery and the post office hold our community together. The New England cracker barrel is gone, but the self-serve coffee and doughnut area in the storeroom behind the meat counter has replaced it as the town meeting place. You don't have to invite the Joneses for a suburban barbecue; you stay in touch with a large number of friends and neighbors

Jeffrey Fowler
and Ron Woodland.

when you get your mail or your milk and eggs. Town government, contractors' daily work plans, real estate deals, and new friendships happen in the store, on the porch, and in the post office.

In August of 1988, I ran into Ron Woodland at the post office. Ron had moved to Worthington and built a small house the same summer that I built my house. We both served on the local Conservation Commission. Ron began to complain that he dreaded returning to his job teaching physics at a regional high school. I complained that the business had grown too large for me to handle. I was burning out. Soon we came to the obvious conclusion that Ron should quit his teaching job and work for Fowler Solar Electric Inc. We met the next day and ironed out a mutually beneficial arrangement for Ron to work part time at my business. Ron worked for me for a year. His employment was a year and a half too early in the chronology of Fowler Solar Electric Inc. Ron went on to start a daycare business with his wife and work part time as a planetarium director. He was happy that the time with me had allowed him to break the professional stagnation of his teaching job. I was happy for Ron's help in my business survival until the arrival of Lea.

After I had paid Ron's wages, I had trouble paying myself the same salary that I had earned before I hired Ron. The business had reached a plateau. We found we had too much labor in the winter. The breakdown of responsibilities was such that I could not take a long unpaid holiday. Ron could not take time off because he needed the income. I was too responsible to cut back Ron's hours and change our agreement. In the past I had never considered doing extensive changes to *Solar Electricity for the Remote Site Home*, because I had always imagined that someone would write a competing work. It never happened. The people who had the skill did not have the time. They were working long hours on their own solar electric businesses. I decided that the time was right for a new book. I had the time to write, Ron wanted to edit, and he wanted to learn computer drawing. He would draw the diagrams for the book.

All winter long, I invested the part of my salary that I was not receiving into labor for the new book. I had faith that the book would later repay me for this investment. We finished the *The Solar Electric Independent Home Book* in the spring. This book was a 200 page 8.5"x11" paperback with many diagrams and photos. A few weeks later the printer delivered our two thousand copies. Ron and I joyfully played like kids with the stacks of real books we had written.

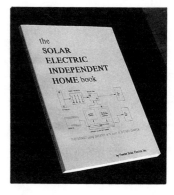

Ron left work in August of 1989. Because we had had too much labor, he had manufactured many extra mounting frames and pre-wired control panels to stay busy. I was well prepared for the fall rush of business. I ran the business by myself until the winter. In January, when the sale of my business to the large distributor fell through, Lea and I dove into the business with a renewed interest. This was my happiest stretch at the business. Lea and I would have daydreaming marketing sessions and play chess games about improving the catalog and reaching the homes that had no power lines. We felt there were thousands of virgin customers in the northeast who had not yet heard of solar electricity. We never liked fighting over customers with other dealers. We did not have the killer instinct to nickel and dime and bad-mouth the competition who were actually our friends.

The business grew that year. Lea was not an engineer, but she was an honest and enthusiastic salesperson. She could explain how we used components every day in our home. I could usually take one day a week away from the telephone to do more relaxing work. I had always been a natural salesman because I never gave any sales pitches. I never pushed equipment because of profit margins or a need to dump it. I used logic and knowledge to accurately portray the pros and cons of any of our products. I tried to make systems cheaper by simplifying the controls. I tried to make our systems easier to install by simplifying the directions. These principles emphasized the role of the buyer as the procurer of knowledge, and my role as the supplier and the provider of knowledge and experience. We always told customers that our advice and experience were free. We would get paid when they finally purchased equipment. Often we had customers who asked questions over years, who then faithfully paid us when they finally purchased solar electric systems. We often spent as much time with the purchaser of a $1,000 system as we did

Photo of Lea and me used in our 1990 catalog.

with purchaser of a $10,000 system. There was good karma in our work.

As the busy summer of 1990 wore on, I began to once again look for a way out of the business. I liked our customers and enjoyed working with them. They were very similar to Lea and me, going through the same decisions that we had made. I was well paid in both dollars and praise. My customers treated me as a friend and mini–folk hero. Lea and I received many standing invitations to visit our customers. I enjoyed each person and each telephone call, but the long days of continuous conversation took their toll. I was exhausted at the end of each day. My energy level was often too intense, and my voice and throat got tired. The long hours of talking irritated my throat. When I caught a cold, it would painfully settle into laryngitis.

In August of 1990, I discussed a potential sale of the business with our accountant, small business consultant, and former Hilltown CDC colleague, Tim Clegg. Tim and I had each grown our respective businesses together since our days at the Hilltown CDC. Tim had also left the bureaucracy for small business independence. Tim felt that our business was healthy and saleable to the right person who could fill the role of owner-manager. Fowler Solar Electric Inc. would appeal to the green political population of western Massachusetts. I had no idea how to advertise or reach prospective buyers. Tim aided me in establishing a fair price. Lea and I made the decision to sell the business, if possible. If not, we would retire the business within two years, and sell off its assets. We would most likely maintain the rights to the book, which we would market with future companion books on our own.

WHEN I WAS A YOUNGSTER, the Glidden estate was one of the nicer properties owned by Worthington summer people. The estate consisted of a large old house, barn, studio, and tennis court on several hundred acres of land. I knew the grandchildren and the children from my days at the general store. The senior Gliddens died, and the heirs sold the property in the early seventies. Several owners later, a group of three families about my age bought the property cooperatively and converted the house into a two-family home and the studio into a summer cottage. Two of the men, Tom Quinn and Steve Schulze, ran one or two small businesses in the upgraded barn. The tennis court became the village volleyball court. Every Sunday, folks in my greater circle of friends gathered there for games.

I had known Tom and Steve for years. I, however, did not know them well. I had recently heard through the grapevine that Tom's and Steve's business was stagnating. I had a brainstorm that they might be two engineers interested in abandoning a dying business and buying my growing business of the future. I called Tom, with whom I was better acquainted. The grapevine gossip was not accurate. Tom explained that Worthington Assembly, which was his business only, was doing well. However, he wanted to bring Steve over to talk with Lea and me. Steve was an electrical engineer who recently had been doing CAD, or computer drafting. His job had ended, and there were no new jobs within commuting distance of Worthington.

Steve Schulze

At our meeting, Tom explained that he was happy owning Worthington Assembly. Steve was looking for a job. He had no money to buy a business, and I did not want an employee. I told Steve to forget all the reasons why this could not work and get back to me if he was truly interested in the business. Steve was very interested. He had worked in alternative energy and wind power. Lea and I decided to stay at the business for the near future if Steve wanted to purchase half of the business. We reached a sale agreement that emphasized sweat equity and an installment purchase. Our plan was to double the sales of the business to pay the salaries of Lea, me, and Steve. If our plans failed, we had contingency plans for ending our agreement. The whole affair was a handshake deal based upon respect and honesty.

WE SHOOK HANDS on our new venture in the middle of August. I worked every available daylight hour for the following six weeks finishing the upstairs of the garage for our new offices. Lea was pregnant. The pregnancy had never felt right to her. One night, Lea woke up with labor pains. We rushed to the hospital, but nothing could be done to prevent the miscarriage. As Lea recovered, I worked long hours to sweat the sorrow of the loss out of my system. I finished the offices on schedule.

Lea has three children from a previous marriage. She also had several miscarriages. She soon wanted to try and have another baby. The miscarriage had been a bitter dose of reality for me. I felt like I had had a baby only to have it taken away. I needed to wait before I could chance it again. When spring came I felt whole again and ready to try. These things take time. Each month Lea's period came on schedule.

June 19, 1991, we left for Rob Roy's place in West Chazy, NY, just south of the Canadian border, to teach a Saturday workshop on solar electricity. It was a difficult trip. We drove the Bronco II, stuffed with our workshop equipment, to our bank to deposit Friday's receipts before we headed for northern New York. This changed our four-hour trip to a five-hour trip. We met hot, smelly, and noisy Friday afternoon traffic on our alternate route. Our five-hour trip became a six-hour trip. I was exhausted from the heat and the driving. We passed a pleasant evening with Rob, Jackie, and their delightful children. Unfortunately, I got caught in one of my common pro-fessional solar electric traps. It was time for Lea and me to go to bed early after our long trip. It was also the prime time of the month for Lea to ovu-late. Rob enjoyed having me, the solar electric expert, as a friend and visitor. I had to go for a tour of his solar electric system that resulted in a long ques-tion and answering session, and a late night to bed after Lea.

Rob was an astronomy and solar solstice enthusiast. His round cord-wood masonry home had orientation points for the solstices. In front of his home he had druid-like placement of stones that related to the passage of the sun. Saturday, the day of the workshop, was the summer solstice. I woke up very early on this longest day of the year. I had had a short but restful night. While Lea and I were relaxing in bed at dawn, we heard Rob wandering around outside. At breakfast Rob told us we had missed the exact moment of the summer solstice. He had gotten up at dawn and watched the light per-form in his system of precisely placed stones.

Several weeks later, Lea snuck out of work before lunch giving me hand signals behind Steve's back while I was speaking on the telephone. She was trying to tell me that she was going to the Worthington Health Clinic for a pregnancy test. I misinterpreted the signals to mean she was going to the general store for Tampax. I was disappointed. I thought another period had come. An hour later, Lea completely surprised me with the news that she was pregnant. After work we bubbled with joy about the pregnancy. We contem-plated the expected date of arrival of the baby. We soon realized that there was only one real possible date of conception. We had conceived our baby exactly at the summer solstice before our solar electric workshop. Our baby was most definitely a solar baby.

On March 7, 1992 Terence Fowler arrived. I feel Terry is the son who was meant for Lea and me, and only Lea and me. I tell him how he had to return after the first pregnancy and miscarriage. I tell him how he was so stubborn and determined even then, that he switched tickets and his place in line, so that he could come to us the second time. Soon after Terry was born, Lea and I made the decision to mostly retire to our homestead. We offered the remaining 50% of the business to Steve. I was an old daddy, and Terry was my first baby. I wanted to spend time with him. Lea and I hoped to share the parenting of our new baby. If all worked well, we would each work part time and maintain a low-expense lifestyle. If we fell behind economically, we would work more after Terry entered school. Steve bought the remainder of the business nine months later.

During our two years with Steve, we tripled the sales of the business at the same time that we increased the profit margins. We expanded our marketing and advertising to reach new customers in the northeast; we doubled the size of the catalog, and we updated the *Solar Electric Independent Home Book* complete with all new computer-drafted diagrams by our new partner. In addition to the book, I wrote and Steve illustrated two companion manuals to the book. Steve took over the bulk of the telephone sales, while Lea and I focused on giving workshops, marketing, advertising, and the vision of the company.

I have never regretted the sale of my business. I enjoy watching the business thrive under Steve's leadership. For years, I felt responsible for all of our customers' solar electric systems. Now I can relax and enjoy my own solar electric system. I look back at Fowler Solar Electric Inc. with pride. I feel that I did environmentally influence the world. We sold 3,000 copies of *Solar Electricity for the Remote Site Home*. To date, Fowler Solar Electric Inc. has sold over 10,000 copies of *The Solar Electric Independent Home Book* and 3,000 copies of the companion manuals.

Terence "Terry" Fowler.

Lea and I love having our homestead to ourselves. When the business was here, we often felt as if we had sold our privacy. Steve and our employees were very considerate of our home and life, but the business was still an intrusion. Several cars drove to our home every day. We had to give tours of our home to customers. The UPS truck raced down our driveway every afternoon. It was not the private country setting we enjoyed before the expansion of the business. In the winter, when the insolation was low, we had to compete with the computers in the office for electricity. Now we are back to the old days and the novelty of a car passing on Bashan Hill Road.

I can see ways my vision of our little company actually influenced the industry. When I wrote the first book, it was difficult to buy and install a solar electric system. I had suffered through my first installation, so I wanted to make it easier for others. If it could be easier, it would ultimately mean more people could jump on the solar bandwagon. All the books I wrote had simple "cookbook" directions for the installation of a solar electric system. Later we were one of the first companies to provide custom installation directions and schematics for all the kits that we sold.

In the early days, solar electric retailers wanted to sell modules and inverters. These were components that came in a box from the distributor and went out in a box in the UPS truck to the customer. Batteries were more difficult to sell. They were heavy and had to be shipped expensively by freight.

The updated *Solar Electric Independent Home Book* and companion manuals.

A mail-order company could not ship batteries over a long distance. The sale of frames, electrical wiring, and components required diagrams and technical assistance. In general, most salespeople at solar electric mail-order companies tried to match prices on modules and inverters to get quick sales. The solar electric home owner was left to make his or her own frames, and to find and install his or her own wiring.

I tried to supply every piece for a solar electric system that the owner wanted to buy from me. I wanted to make it possible for the novice to have a solar electric system. I designed prewired charge-controller panels. These panels used the simplest switching gear, at the same time they utilized UL-listed components and satisfied the electrical code. The homeowner only needed to screw the panel to the wall, send the module wiring into a circuit-breaker box, and attach the prewired ten-foot cable to the battery bank. The homeowner did not have to try to learn to be an electrician. A handful of other solar electric pioneers followed similar approaches. Over time, we forced most of the mail-order companies to provide much more than the quick discount sale.

In my first book, I logically wrote a chapter on how to size a solar electric system. I finished the chapter with an innovative "seat of the pants" approach to system sizing. I provided the reader with detailed descriptions of five potential kits. I estimated the output of each kit and described the kinds of loads that could be powered and for how many hours per day in both the summer and the low sun of winter. Each kit listed all of the components necessary for an installation, the average prices for the components, and a price for the kit as a whole. I worked under the assumption that the potential buyer could not, and did not want to, design a solar electric system for his or her home. The buyer needed to see what a system would do and know what that performance would cost. People could not afford to purchase a solar electric system to meet the load of all the appliances they wanted to run. They needed to see what system fit into their budget, and what that system could do for them.

The kit approach to solar electric system sizing aided me in my business. Customers bought and read the book before they purchased a system. I paralleled my catalog to the kits in the book. The reader of the book would call to order a kit in the book, one that was also sold in the catalog. My books were sold by other solar electric businesses across the country. Soon I noticed that Windy Dankoff and others added a sales line to their catalog recommendation of my book. "We can supply you with the kits in this book." Some businesses were threatened by a book written by a potential competitor. My business primarily served the Northeast. In reality, I simply had a vision. I wanted to get more, well-designed, and complete solar electric systems into the hands of the independent people. I was giving away my ideas of how to supply customers. I was providing templates for other businesses to copy. I had little interest in national sales.

My second book, *The Solar Electric Independent Home Book*, had a much expanded section on kits. Each sample kit was accompanied by a detailed schematic for its wiring and grounding. The book also had a detailed chapter on installation that applied to the installation of any of the kits. I would guess that most small dealers in the country owned a copy of this book. This book supplied system design and installation instructions for many beginning small dealers. We chose to publish much of our expertise and make it available to both the end users and anyone who wanted to start his or her own solar electric business.

I have given solar electric workshops all over the northeast. I have seen many solar electric installations. Many solar electric systems have wiring that most resembles a rat's nest. Many other systems have no balance between the different components in the system; the battery bank is too small, or the inverter is too large. The imbalance in these systems creates a solar electric system that does not perform well for the dollars spent. I carefully analyze the independent home systems in the latest books and the latest articles. I take great pride that in my years at Fowler Solar Electric Inc., I focused on providing quality, complete kits and directions that resulted in excellent solar electric systems, installed by homeowners. Each good installation that we helped someone accomplish has provided a good example for the next installation of the friend or neighbor down the road.

CHAPTER 24

Appliances and Energy Conservation

B EFORE THE MOVE to Bashan Hill, I lived in a small
house with very modest electrical demands. My large
loads consisted of a well pump, a washing machine, a
freezer, and a refrigerator. My monthly electric bill
averaged 250 kilowatt-hours. I worked hard to keep
my usage at this level. I conscientiously turned off lights that were
not in use. In the mold of a frugal rural New Englander, the outside
light was turned off by 9 P.M. We have always joked in Worthington that you can locate all the non-locals by driving home late at
night. The non-locals live in the houses that leave their outside lights
on all night. We guessed that former city folk could not sleep in the
country without the security of a simulated streetlight.

I do not have a monthly electric bill here on Bashan Hill. My
solar electric system has no kilowatt-hour meter. During the best
solar time of summer, my current solar electric system is capable of
producing 150 kilowatt-hours of electrical energy per month.
During the worst solar time of winter, the solar electric system produces up to 75 kilowatt-hours per month. In the winter, we may be
able to use the majority of the 75 kilowatt-hours produced. In the
summer, much of the electrical energy produced by the modules is
wasted because the batteries reach a state of full charge early in the
day, and the modules are effectively turned off at this point. My current home, though it survives on less than half of the electrical energy that was needed to power my former home, provides us with an
almost identical level of comfort.

We use much less electrical energy in our solar electric home than

I did in my previous home, because we have learned about our loads and appliances. Conservation of electrical energy is a part of our lives, as it was a part of our home design. I have always told solar electric customers that alternative energy works well, not because it can simulate the production of the local utility company, but because the owner of an alternative energy home learns to get the same results on less than half of the electrical energy. Some of this reduction in electrical energy use is a product of energy conservation through the selection of extremely efficient appliances. The rest of the reduction results from replacing electrical appliances with ones powered by other sources of energy such as wood, propane, or thermal solar energy.

Solar electricity became cost effective for our home, as it does for most remote-site homes, because of the extreme cost of a power-line extension. In our case, we would have had to pay the utility company $20,000, in today's dollars, for the line extension. This sum would have had to be paid before the utility company could start work on the line extension. Our eventual solar electric system only cost about $12,000. We theoretically saved $8,000. The money was saved because we chose not to spend the full $20,000, and we chose to employ electrical energy conservation.

People in our area who buy 250 kilowatt-hours of electrical energy per month from the local utility company pay about 10 cents per kilowatt-hour. Every month they write out a check for approximately $25 to Northeast Utilities. We had to purchase all of the components for our solar electric system. Each part has a certain life before it must be replaced or overhauled. When we take into consideration the costs of all the parts and all their maintenance over a long period of time (for example ten years) and then divide it by the total number of kilowatt-hours that is produced over the same period of time, we find that we pay about 30 cents per kilowatt-hour for our electrical energy.

I have my own methods of comparing the cost of our solar electricity on Bashan Hill to the cost I would pay to the utility company. The logical part of me approaches the comparison by dollars. I spent $8,000 less to get electricity to my home. I also produce about 100 kilowatt-hours of electricity per month for ten years for free. This has a value of $1,200 at 10 cents per kilowatt-hour. I have saved $8,000 plus $1,200, or $9,200. After ten years (the average life of the components in a solar electric system) I will pay 20 cents per kilowatt hour more than I would have paid on the grid. I stay ahead of the cost from the utility company until thirty-eight years after my ten years of initial free electricity. My personal way of looking at the savings of my solar electric system is very basic. I saved $8,000 originally and got ten years of free electricity in the process. After the ten years, when I start to maintain the system, I will begin to pay 30 cents per kilowatt hour for 100 kilowatt hours per month. I will pay $30 per month for electricity. If I were on the grid, my family and I would soon use more electricity and spend $50 per month.

"The lack of a power line is a poor real estate investment," friends have pointed out. "Your home is worth $20,000 less." I reply, "I do not have to spend money on more land to ward off the pressure of development. My home is not for sale. It has no dollar value."

There are numerous ways to compare the cost of a solar electric system to the cost of electrical energy from the power company. It is obvious that we pay three times as much for solar electricity as we would for the electricity from the power company. Anyone can see that someone would seem foolish to replace 10-cent-per-kilowatt-hour electricity with 30-cent-per-kilowatt-hour electricity if a power line were already running into his or her home. But what if everyone did? Perhaps it would be a good deal.

Today, people attached to the grid pay 10 cents per kilowatt-hour directly to the utility company. Additionally, through their tax dollars, they fund the Nuclear Regulatory Commission, subsidies for nuclear power plants, the federal Department of Energy, the state Department of Public Utilities, subsidies for manufacturers' use of electricity, programs to find more oil and coal, and long distance power line projects. The list goes on. Additionally, there are the negative costs of electrical production by environmentally unsound means. Who pays to decommission a nuclear power plant? What are the costs of rebuilding a bridge that has had a shortened life due to acid rain? Who pays for the higher cost of food because of a decreased harvest due to acid rain? What is the cost for medical insurance for treating ailments that result from poor air quality? Many of us feel that the electrical energy from the power company is deceptively more expensive than the 30 cents per kilowatt-hour that we pay for solar electricity.

In 1973, there was the great energy crisis. People in New England bought up all the wood stoves available. They switched to wood heat if they had electric heat, but also if they heated with oil or gas. The first consideration was the inflated energy costs due to the shortage. When the energy crisis subsided, many people never gave up their wood stoves. Wood heat is not a great deal. Most anyone would do better to work the same number of extra Saturdays at his or her job than to cut, split, and stack wood. Why do they still do it? I think they want their independence. They want control over their own welfare and survival. As recently as the 1930s, rural American homes were completely independent. They heated with wood and lighted the house with kerosene lamps. There was no refrigeration, and gardens were a necessity. I feel that the return to wood stoves is an indication that some people miss some of that self-sufficiency. I strongly feel there may be a groundswell to independent power production when the cost of independently produced electrical power falls to the same price as power from the power company. Many independent people in my part of the country may simply say, "I would rather do it myself." Once they do it for themselves at the same price, they will invest a lot of personal energy in conservation and gloat that they are really doing it for less.

The most committed environmentalists conserve electrical energy in their homes to lessen the amount of pollution. Industries reduce their consumption of electrical energy for economic reasons. Average homeowners have no idea what an appliance in their home costs to run. They have little incentive to purchase a more efficient appliance, because a more efficient appliance initially costs more. The more efficient appliance will result in little savings on the monthly bill because the homeowners pay only about 10 cents per kilowatt-hour. Perhaps if they were paying directly to the power company the true cost for the production of electrical energy, they would feel different.

Solar electric homeowners pay 30 cents per kilowatt-hour for electricity. Conservation of electrical energy results in a large savings at this higher price per kilowatt-hour. Furthermore, solar electric homeowners must pay for their electrical energy use in advance rather than a little bit each month. If a solar electric homeowner wants more electricity, he or she must make a capital investment in a larger solar electric system that will produce the greater amount of electricity for the next ten to twenty years. For these reasons, it is economically imperative for a solar electric homeowner to make the most efficient use of electrical energy.

In our solar electric home, we have chosen to replace many large electric appliances with appliances that can be efficiently powered by other forms of energy. Our house is heated by passive solar energy and wood heat. We need no electric pumps or fans to circulate heat. As in most rural homes, we use propane to cook and heat water. We have replaced our conventional propane hot-water heater with a super-efficient tankless, or on-demand, propane hot-water heater. Our refrigerator is also powered by propane. A conventional refrigerator powered by our 30-cent-per-kilowatt-hour electricity would be much less economical than our propane refrigerator.

Our Sibir ten-cubic-foot propane refrigerator.

For the first four years that I lived on Bashan Hill Road, I did not own a refrigerator. I had bought and sold my giant pink propane refrigerator before I even hooked it up. Eventually Bob and Karen Cook gave me an old Servel gas refrigerator, a repair manual, and enough extra parts for me to get it working. The Servel used a lot of propane. The large amount of inefficiently burned propane also produced a large amount of fumes. I had always been able to smell the burned gases at the Cooks' when I visited their kitchen. Because my home was a much tighter than theirs, I did not dare to install the refrigerator in my kitchen. I kept the Servel in the basement at the bottom of the cellar stairs. The cool environment meant that the refrigerator burned less propane and made less pollution. The natural air changes in the cellar kept the pollution out of my living and breathing space above.

Rick and Sharyn had purchased Big Pink, the giant refrigerator. They used it for a few critical months and passed it on for the same $100 that I and they had paid. That refrigerator was a gas hog. Luckily, they understood that I had innocently sold it to them. A few years later I sold them a solar

Our Aquastar tankless, propane hot water heater.

electric system. Soon Rick came to me to buy a new propane refrigerator. He knew that I did not list them in my catalog. Still, he insisted that he wanted to buy it from me because I was his friend and alternative energy dealer. Rick's mom had given Rick and Sharyn money to buy a refrigerator as a wedding present. I researched gas refrigerators. The best gas refrigerator was manufactured in Sweden by Sibir. I sold one to Rick and Sharyn for $900. I felt so guilty about the exorbitant price that I sold it to them at $25 above my cost. I could not believe that they had spent so much.

I saw Rick and Sharyn often. They loved their refrigerator. Soon they drove over to give me a copy of Lehman's Non-Electric Catalog, which featured Sibir refrigerators. I called Jay Lehman and found that he was the sole importer of Sibirs sold east of the Mississippi. My business sold a few refrigerators. Finally, I purchased a display model and installed it in my own kitchen. That Sibir is still there. It has performed flawlessly for eight years. This true deep freezer keeps food at −10° F. We use only five dollars' worth of propane per month.

My first hot water heater was a used standard gas hot-water heater with a 30 gallon tank. This hot-water heater performed adequately after I added additional insulation. Four years later, I made a group purchase of Aquastar tankless, or on-demand, hot-water heaters for some friends. One friend returned his heater and wanted his money back. Unfortunately, I could not return the heater to the distributor. I became the instant owner of an expensive hot-water heater. Soon, Jonathan plumbed the new water heater into my home. The Aquastar was an excellent and efficient heater. I instantly cut my propane use in half.

I have owned many propane stoves. I like to joke that I bought used stoves and threw them away when they needed their ovens cleaned. Eight years ago I was finally ready to buy a new stove that would fit in my kitchen and into its design. I soon found that all of the stoves on display at the appliance store used electronic ignitions. Pilots were outdated. The amount of electricity was small. The problem was that the inverter had to be on to power the stove. If the stove kept the inverter on all of the time, it would spoil the efficiency of the inverter. If the electronic ignition did not turn the inverter on, then another light would have to be on during the operation of the stove to keep the inverter running.

I finally found that there were gas stoves that still had pilots and no electronic ignitions. These stoves were always the basic model in the product line of any stove manufacturer. These basic stoves were white only. I wanted a stainless steel or black stove. I wanted the look of a commercial restaurant stove in my kitchen. One day I found a black propane stove with no electronic ignition in a small appliance store. Ironically, the stove was manufactured by General Electric Co. Unfortunately, there were no options available on this basic model. I had to settle for its brushed chrome top that only looked like stainless steel. I could not get a see-through oven door. It was still better than a plain old white stove.

Water pumping was probably the largest load in my home. I had only one option for a water pump. Because I had a deep well, I needed a submersible pump. The static level of the water in the well was too low to use a centrifugal pump. I chose the highest quality, smallest 120V submersible pump available. More efficient DC pumps evolved in the alternative energy market over the years since I installed my pump. The best ones supplied the same amount of water that mine supplied for one-half of the electrical energy. Unfortunately, none were compatible with my water system. Most were centrifugal pumps or diaphragm pumps that could not draw water up from the depth of the water in my well. The DC submersible pumps were too small, and they ran at low voltage. My well was 200 feet from my home, and the line losses of low voltage electricity over the long distance would create a great inefficiency.

My alternative energy home evolved around the evolution of the inverter. The first inverters had not been tested to run the appliances that I wanted to run. There was no information on inverters and appliances, and little data available from the inverter companies. I researched as best I could and experimented the rest of the way. My first problem load for the inverter was the deep-well pump that required a high surge to start under load. Soon I wanted to add a washing machine to my home. Some alternative energy users, before the availability of inverters, had converted old wringer washing machines to run on low voltage DC. I wanted a conventional washing machine. I believed the alternative energy world had to offer more normal appliances if it was to grow.

Washing machines were difficult loads for inverters because they required a surge of electricity to start the motion of the tumbler. The tumbler was loaded with water and clothes. The washer had to move this load from rest to motion. Soon the tumbler direction was reversed and the motor again was under great load. My research determined that a heavy-duty, gear-driven washer quickly sent an inverter into overload. Standard washers, which utilized a clutch and belt drive, could be powered by an inverter as long as the washer had a capacitor-start motor and not a split-phase motor. A split-phase motor requires a large slug of electricity to start rotating. A capacitor-start motor adds a capacitor to the starting windings of a split-phase motor. The slug of current that was needed to start the motor is evened out by the capacitor. Effectively, the inverter only needs to supply a medium surge of current for a medium period of time. In the absence of the capacitor, the inverter tried to supply a large amount of electricity all at once, and went into overload and shut itself off.

I purchased my first washing machine at a large appliance store in Pittsfield. I chose a General Electric washer that had an excellent Energy

Our propane stove without an electronic ignition.

Our kitchen.

Guide rating. The first salesman knew nothing about the actual workings of the washing machine. He only sold appliances. The second salesman appeared to know more. He assured me that the washer had a capacitor-start motor. I installed the washer the next day. I turned it on and watched it quickly pull the inverter into overload. When I called the service department, they quickly informed me that the motor was only a split-phase motor. All the research in the world fails when the product is misrepresented. The serviceman suggested that I purchase a $10 low-voltage device sold to solve similar problems. The device turned out to be a capacitor to be installed in the starting winding circuit to convert the split-phase motor to a capacitor-start motor. The experience actually had great value. I recorded the micro-farad rating of the capacitor. I later advised customers how to add a capacitor. I supplied capacitors to them for $5.

My first inverters powered my well pump, and they powered the washing machine, but they would not power both of them together. Under these conditions, my automatic washing machine was less than automatic. To wash clothes I had to pump up my large water tank and turn off the pump before I started washing. Then I turned on the washing machine. In the middle of the load I had to stop the washing machine, turn on the pump to refill the tank, turn the pump off, and then turn the washing machine on for the rest of its cycles. Trace inverters were a true breakthrough for my home and my customers' alternative energy homes. The Trace 2000 watt 24V inverter had such a high surge capability that it could start both my washing machine and my well pump at the same time. My alternative energy home was almost automatic. All I needed to remember was to operate no other large loads, such as a large vacuum cleaner or a circular saw, while the wash was on.

In 1985 I started searching for my first computer. My first and most obvious question for the salespeople at the computer stores was, "Will the computer run on an inverter? The inverter does not produce true sine wave." The first answer was always, "It will void the warranty." I finally called several small alternative energy companies in California. The small companies were often in homes of people who lived with alternative energy systems. They all said, "So far so good." I bought my first computer from Clark at Infinity Computer, a maker and seller of IBM clones. Clark was an engineer from Tufts, my alma mater. He was not a salesperson, but a computer technician. His response was, "If it doesn't work, I'll take it back. The power supply doesn't need a true sine wave. A computer will operate on a UPS (uninterrupted power supply) system that contains a small square-wave inverter. Those salesmen don't know anything." That original computer is still running at Ron Woodland's home today.

My first solar electric system's most important function was to provide light for my house. I chose each light in my home very carefully. If I could utilize a focused 25 watt reading light instead of a 50 watt general lighting fixture, then I could power the smaller light for more hours per day. I had a finite amount of electrical energy available from my small solar electric system. I needed to place the most efficient lights in the best locations to get the most out of my finite electrical budget. These first lights were all powered by my 12V battery bank. There were no highly efficient 12V fluorescent lights at the time. However, the 12V incandescent bulbs that I did use were twice as efficient as their 120V counterparts. The lower voltage and higher amperage 12V incandescent bulbs operated in a better efficiency range for the tungsten filaments. These 12V bulbs produced more light and less heat per watt.

I carefully selected fixtures and lamps and thought about their placement. For reading, I used inexpensive drafting lights that concentrate a cone of light on a desk or a book. For general lighting, I selected clear glass fixtures or highly transparent shades. I wanted all of the lumens produced by the light bulb to pass to the room. My house was a bright house. The relatively small amount of light reflected off the many light colored or white surfaces. It amazes me even today when I enter a gridpowered house that has lights hidden in recessed fixtures or under almost opaque lamp shades. These houses will often use 400 to 600 watts to light a single room that I would be able to light for 40 to 60 watts.

As my solar electric system expanded and an inverter allowed me to utilize the 120VAC circuits of my house, I added 120VAC lights. The hall lights were small 25 watt 120VAC incandescent bulbs. These lights needed only to light the hallway for a brief period of time as someone passed through at night. In the downstairs rooms, I installed ceiling fixtures that looked like old gaslight fixtures. The fixtures consisted of brass and clear glass. I fitted each of these fixtures with a soft-white 120VAC 40 watt incandescent bulb.

The small soft-white bulb created no glare at the same time that the glass allowed all of the light from the bulb to illuminate the room.

I have never cared for fluorescent lighting. I have always been irritated by its odd spectrum of light and 60 cycle flicker. The few shop fluorescent lights that I tried, amplified the 60 cycle electrical noise of the quasi-sine-wave inverter into a menacing background hum. Lighting companies began to manufacture compact fluorescent lights that were made to be screwed into standard bulb sockets. The internal ballasts of these lights ran at a high frequency, thus eliminating all flicker and 60 cycle hum. The first generations came in sizes that were inconvenient to install.

AROUND 1989 Osram Corporation began distributing a new design of compact fluorescent bulbs. This changed and greatly enhanced the efficiency of the lighting in my home. The bulbs produced a spectrum closer to daylight and fit into almost the same space as a standard incandescent bulb. The lights came on instantly with no irritating flicker flicker flicker. Unlike previous models of compact fluorescent bulbs, these bulbs would turn on the inverter from its sensing mode. The most popular bulb required 15 watts. It was rated to produce the same amount of light as a 60 watt 120VAC incandescent bulb. I found this rating was for initial light output. The bulb, when allowed to run for more than five minutes, actually produced the same amount of light as a 75 watt 120VAC incandescent bulb.

My current lighting system is a hybrid of all the earlier stages of lighting design. I have one low-voltage DC lamp or light fixture in each room of the house. A few of these lights are 24VDC fluorescent lights. All these lights are in useful locations. Some of them are bedroom reading lights that we use as primary lights; others we only use when the inverter is turned off during severe lightning storms. Half of our home lights use standard 120VAC incandescent bulbs that are carefully placed in very efficient fixtures, or relegated to very short-time use, as in a hallway. The remaining house lights are fitted with 11 watt, 15 watt, or 20 watt compact fluorescent bulbs. These lights tend to be inside translucent shades that increase the feeling of natural light. The cellar and garage are lit by bare-bulb porcelain fixtures that are fitted with one or two 20 watt compact fluorescent bulbs for maximum light and minimum electrical use.

Today, a new alternative energy home does not have to suffer through the experimentation that I did. There are books and catalogs full of applicable information. One still needs to carefully consider all appliances and lights. We pay 30 cents per kilowatt-hour for our electricity up front for the next ten or twenty years. We need to be able to get the best performance out of all loads to insure that we can conserve our way through the poor insolation times of the year. We would have a cleaner world if all households in the world played the same games that alternative energy users must play.

A compact fluorescent bulb.

CHAPTER 25

The Homestead and the Alternative Energy System

OUR HOMESTEAD has evolved around the capabilities of our alternative energy system, and our alternative energy system has evolved around the needs of our homestead. These two systems are always dependent upon one another. An improvement in our homestead or lifestyle that requires electrical energy can only be undertaken if our alternative energy system has sufficient output at the appropriate times of year to power the new loads, or if we are willing to expand our alternative energy system. We try not to increase our need for electrical energy so we don't have to expand our alternative energy system. Often this means we choose a project and transform it from its conventional design to a more efficient one that utilizes little or no electrical energy. Sometimes a new undertaking is designed around a seasonal surplus.

We have a ten-cubic-foot propane refrigerator; one and a half cubic feet of this space is a deep freezer. This refrigerator is much smaller than the 15–20 cubic foot models that most American homes have. Most grid homes that have large gardens, as we do, also have a 15–20 cubic-foot freezer to store vegetables and fruit for the winter. Standard refrigerators and freezers are underinsulated and extremely inefficient appliances. Our solar electric system could not afford them. We have developed our homestead in different ways to decrease our emphasis on refrigeration.

My first choice for food storage came before I had even started construction on my house. I planted a garden. The garden supplied fresh vegetables for four months of the year. I picked these vegeta-

The inside of our cold
storage room.

bles and immediately ate them. Without the garden I would have had to buy
my vegetables once a week at the supermarket and refrigerate them until I ate
them. I had no refrigeration at all for the first four years at my home. I
learned to plant some cool weather vegetables, such as kale and fall lettuce,
which produced until snow cover in December. Additionally, I learned to
plant extra root crops such as potatoes, carrots, and parsnips that could stay
underground and be harvested as I needed them until the ground froze in
November. The extension of my garden year gave me vegetables for two
extra months. Some of these vegetables were stored for an additional time in
my cool basement.

After four years without refrigeration, I got the old Servel gas refrigerator
that was soon followed by the new Sibir gas refrigerator. The small Sibir
refrigerator was large for my one-person household. I froze vegetables from
the garden and filled the small deep freezer. Between the late fall vegetables
and the frozen vegetables, I had vegetables through the winter. With the
arrival of Lea and her children, and then our son Terry, the needs of my single person household greatly expanded. The freezer was too small for the
quantity of frozen vegetables needed to feed a family. The kids consumed
large quantities of milk and other perishables. The propane refrigerator was
full. We had to consider the small size of the refrigerator when we shopped.

FOUR YEARS AGO, Lea and I began to make plans for more cold storage.
My friends Richard and Meg showed me their root cellar before I built my
house. I had planned to build my own root cellar. For years I just never
found the time. We considered a low-voltage DC efficient refrigerator
and/or freezer, an additional gas refrigerator for the cellar, a root cellar, a

conventional freezer to be located at a friend's house, or some combination of the above.

We strongly considered a very expensive, efficient 24VDC refrigerator. The large model was several inches deeper than a standard kitchen counter. We could not fit one in the space available in our kitchen. Furthermore, we did not want the extra electrical load. The refrigerator could not be shut off in times of low insolation. We had had several customers who had unbalanced their winter electrical energy budget when they added the everyday load of the electric refrigerator. During the several weeks of no sun in December, the consistent electrical demand of the refrigerator was too much. We were less inclined to purchase a 24VDC freezer, because the need to keep the food colder would result in twice the consumption of electrical energy of a 24VDC refrigerator.

We decided not to spend $1,100 on an additional gas refrigerator and pay an extra five dollars a month for propane. The small freezer would not furnish us with enough extra space for frozen vegetables. Because we had fresh vegetables in the summer, our greatest need for extra refrigeration was in the non-summer months. This balance seemed to place the choice of a root cellar above the expensive gas refrigerator. We designed and built more of a cold storage area than a true root cellar. In our cellar we had an eight-foot by twelve-foot area partially enclosed by the concrete walls that supported the suspended slab above it. We insulated all the walls between the cellar and the room. We left the walls facing the cold ground outside of the cellar uninsulated. We insulated the ceiling of the cold storage room from the concrete slab above that was the floor of the heated kitchen.

In the depth of winter, the cellar had always maintained a temperature of 45° F. Since this room was insulated from the cellar, it stayed colder than the 45° F. I installed a small 24VDC fan to bring in cold air to cool the area to about 38° F. In fall and spring, I ran the fan on cold nights to keep the room as cold as possible, though it never dropped as low as 38° F. This room was more of a cold storage area and less of a root cellar because it did not have a dirt floor. We installed a series of shelves to store foods that otherwise would have been stored in a large refrigerator. These shelves were large to accommodate foods bought in bulk such as butter, cheese, and large bags of vegetables or fruits. Large batches of soups, lasagna, or other leftovers easily fit on shelves, where they would never have fit in a conventional refrigerator. The root-crop vegetables were packed in sawdust or sand in containers on the floor. Apples and onions were stored in buckets or boxes.

During the summer, the room gradually increased in temperature from 50° F. in May to 60° F. in August. The room was still cold enough to store many foods for shorter periods of time. In the cooler months we loved the area as a method of expansion of our small Sibir refrigerator. We could cook large batches of foods and leave them in their large pots or pans. We used the area to cool down warm foods before we put them in the Sibir refrigerator or freezer. Lea and I feel this area did not function as well as it should have for storing specific root-cellar crops, which need different ranges of humidity and temperature. Many of the deficiencies of performance were due to our lack of experience. We gradually learned to do a better job storing the vegetables and monitoring the conditions of the root cellar.

My first garden at Bashan Hill was terrible. The ground had been poorly tilled. As the summer progressed, I let it overgrow with weeds and grass as my free time decreased. It still provided me with my summer vegetables. The first garden taught me that the garden's efficiency would be directly related to my alternative energy usage. That first summer was very dry. To have an above-average garden I needed to painstakingly water it. Most New England gardens survive with mostly normal rainfall. I wanted a more productive garden. I needed to water it at precisely the right time to have seeds germinate quickly, to make seedlings thrive in dry weather, and to have vegetables grow to their maximum size. That first year I had to haul water until my pump was installed. Until I bought an inverter, I had to run a generator to pump the water for the garden. Even with later larger solar electric systems, I had to budget the pumping and watering.

During the latter part of my first winter, I planned my next garden. I read some books on raised-bed gardening. This whole method of gardening appealed to me. Some aspects were very well suited to my alternative energy homestead. The raised-bed design reduces the area required to grow vegetables, or stated differently, higher yields can be obtained from a small surface area. The beds are cultivated very deeply, often two feet. Plants can be grown more closely spaced together because their roots grow downward in the deeply cultivated soil, instead of growing outward in a shallow area around the plant. The decreased surface area means a smaller area to water, so less water is required. The lesser surface area reduces evaporation of water from the soil on dry days. For my garden the reduced demand for water would result in less pumping of water and less generator run time or demand on my solar electric system.

I designed my raised bed garden that first winter. I also ordered a large rototiller. I planned to till and rehill my raised beds for several years. I planned to move the garden to the south side of my house and abandon the first garden. A surprise nor'easter spoiled my spring schedule. Next, a farm accident prevented a local man from clearing my tree stumps. I abandoned my plans to move my garden.

My first garden had been poorly tilled. The plants grew between the boulders and the stubborn roots of the apple tree saplings I had cleared. This second year, I tilled the garden six times in all directions. I dug out boulders and roots. When the tilling was completed, I hilled the area into six raised beds, each four feet wide and 20 feet long. The earth in the beds was tilled eight inches down. The soil hilled on the top of the beds added another eight inches. This resulted in raised beds with sixteen inches of cultivated topsoil. A two-foot wide path separated each bed from the next. For the next several years I rototilled and hilled these beds. Each season I added organic matter and cow manure. The beds developed rich, friable soil. At that point, I sold the rototiller and turned over the beds each spring quite easily by hand.

Lea greatly changed my garden. I had always been motivated to get my garden off to an early start. As summer progressed other work became more important. The garden was left to survive on its own. I reasoned that the garden produced too much food, so there was no real need to baby the garden all summer only to give away the extra vegetables. Lea arrived in June. She was a world-class long-distance runner, whose appetite created a need for additional food from the garden. Lea weeded and maintained the garden to keep it productive through the fall.

The garden needed to be more productive than it had been in the past. It had to feed two hearty eaters and three children who came part-time. I analyzed our garden and our garden labor that winter. Lea's strengths were giving tender loving care to the plants and keeping the garden weed-free. My strengths were experience in gardening in our climate, getting a garden started early, and designing a garden system. I decided to upgrade our garden design to create a less labor-intensive garden that Lea could tend. Each spring I could use my muscle for the bull work of turning the soil and planting. During the summer, Lea could weed and care for the plants.

My garden had always been productive, but I had always analyzed it and recorded its deficiencies. Raised beds grow plants so close together that few weeds have room to grow. After the early weedings, when the plants are young, little weeding is necessary because the plants outcompete the weeds. However, the two-foot space between the beds in my garden grew weeds and grasses that invaded the beds from their perimeters. It was necessary to weed these areas. After Rolo died, it became necessary to install a small electric fence around the garden to keep the rabbits out. The area under this fence also had to be weeded to keep weeds from touching the wires and grounding out the fence.

Our garden required minimal watering because of the small surface area of the beds. I read product information on underground leaker-hose water systems. These systems use less water than conventional surface watering, which causes much of the water to evaporate. It is difficult for water to percolate into rich humus content soil. Consequently, surface watering keeps the top of the soil wet. Plants are encouraged to grow lateral roots near the surface to obtain moisture, instead of growing deep roots. Underground watering eliminates evaporation. The plants' roots grow downward to the water supply. Secondly, plants should not be surface watered during the middle of the day. The droplets that bead up on the surface of leaves act like small magnifying

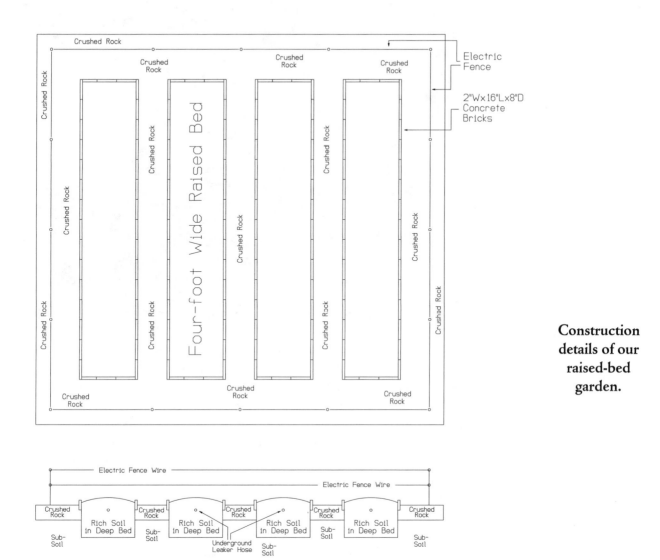

Construction details of our raised-bed garden.

glasses, causing the strong sun to burn spots on the leaves. Watering is best done in the evening, when the cooler temperatures cause less evaporation and allow the water to percolate into the soil. An hour of surface watering each night adds significantly to the garden labor budget. Underground watering is labor free. The garden can be watered very slowly and consistently for long periods of time.

The next spring, Lea and I performed a major overhaul of our raised-bed garden. We worked very hard to set up a system that would eliminate the bulk of the garden labor in the future. I own a post-hole digger that attaches to the PTO on my tractor. This giant screw-type mechanism can drill a hole

Our finished
raised bed garden.

in the ground one foot in diameter and four feet deep. The deeper the cultivated soil is in a raised bed, the better. I drove the tractor up and down the
garden, straddling each raised bed, digging as many three-foot deep holes as
possible. I added manure and refilled the holes. I mostly managed to turn
over the soil to a depth of three feet. Next, I hilled the beds with the twelve
inches of topsoil that I removed from the former paths surrounding the beds.
The beds were now six inches higher because of the addition of the soil dug
out of these areas.

Leaker-hose systems for gardens had been composed of expensive special-
ty hosing. At this time, leaker hose had just become inexpensive and fitted
with standard garden hose connectors. I installed leaker hose at a depth of
eight to ten inches in all of the beds. The watering system was zoned such
that all of the garden or only parts of the garden could be watered. I filled all
of the deep ditches between and around the raised beds with three-quarter-
inch crushed rock. To separate the edges of each bed from the surrounding
crushed rock, I installed 8"x16"x2" patio bricks with the sixteen-inch sides
end-to-end, and the eight-inch dimension up, to create a small curbing. This
would keep the soil out of the crushed rock path and the crushed rock out of
the beds. Over the crushed rock path that surrounded the garden, I installed
the electric fence.

Lea has gradually improved the landscaping around our home. I try to
keep her happy and design all projects to both meet the parameters of the project and enhance the appearance of our homestead. We built a vegetable garden that was also beautiful to look at. Our finished garden was neat and efficient. The messy sprawl of weeds and dirt between the beds was replaced with

neat crushed rock paths. The electric fence that was installed over the crushed rock perimeter path was no longer threatened by weeds touching the wire. The paths, which were functionally ditches filled with crushed rock, drained the garden in the wet early spring, aiding early cultivation of the beds. The underground water system efficiently provided water to the roots of the plants with no labor from me or Lea. Each spring, I turn over each bed easily with a manure fork, and we plant. At that point, the bulk of the hard labor is finished for the year. For the remainder of the growing season Lea weeds the small amount of weeds in the beds and replants appropriate crops.

Two years ago, when we sold Fowler Solar Electric Inc., we became very enthusiastic about expanding the self-sufficiency of our homestead. We planted some fruit trees, planted an asparagus bed, and added space to our garden. We bought a conventional sixteen-cubic-foot freezer and installed it in the basement of Mrs. Joslyn's house. For two growing seasons, we filled the freezer with fruits and vegetables. Also during this time, we improved my design of polyethylene cloches for extending the growing season of our garden. We have decided it is too hard to freeze vegetables and transport them back and forth to our freezer at Mrs. Joslyn's. We will soon sell that freezer. We would love to have a freezer in our own basement, but we are not happy with available 24VDC models or gas models. For the immediate future, we will live without one. We have easily managed to grow lettuce until December under cloches in our garden. Next year we hope to extend that further and have lettuce ready to eat in April. We will grow additional vegetables that are suited for winter storage. We are fortunate to have a supermarket-sized health food store in our area which sells organic produce in the months when we cannot grow it.

A year ago, I began to suffer from a chronic nasal or sinus irritation. My doctor recommended many sensible ways to minimize the irritation. The major home recommendation was to eliminate the wood stoves, which create a dry climate in a house that irritates nasal and sinus membranes. Additionally, wood stoves leak smoke and ash. My final solution was to sell our Glenwood cookstove and our Vermont Castings wood stove. We could not heat with a gas furnace because of the electrical requirements of the circulating pumps and fans. I chose to install a large modern wood stove with a catalytic converter in the basement. I wanted to continue to heat with wood, but at the same time remove the irritation of the wood heat from the living quarters of the home. Because our home and our basement were well insulated, I hoped to heat it with no registers, ducts, or circulating fans, and simply allow the rising heat from the insulated basement to heat the house. Previously we had needed only two cords of wood to heat our home. We would probably burn more wood heating from the basement, but the total amount would still be less than three cords.

Our basement heat worked well last winter. The house had considerably less dust and ash, and the air was not dry. My ailment gradually improved. The house felt much warmer at a lower temperature because the heat came from the floor. The suspended concrete slabs were heated from below. They became heat sinks that evened out the temperature of the house. One room of the house did not heat well. The kitchen concrete floor slab was insulated from the basement heat of the wood stove by the insulated root cellar directly underneath it. The heated slabs in the other rooms have worked so well that we have decided to move our cold storage room to the north side of the cellar.

We have an additional reason for moving the cold storage room. The cellar heat created a problem. Our cold storage room, on average, became a little warmer than it had been when the unheated cellar temperature had been 45° F. in the winter. The new cold storage room will be more heavily insulated to compensate for the warmer cellar. It is always a shame to redo a project like a cold storage room. In this case the benefits of the cellar floor heat far outweigh materials and labor involved in the move. Our new heating requires no loads on our electrical system. I will happily move the cold storage area. After experimenting for several years with our cold storage area we have better ideas of what is needed, so we will build a smaller and much improved version.

Our heated basement has benefited us in many other ways. The warm, dry area has become our winter clothes dryer. We have ceiling-mounted clotheslines and folding clothes racks on which to dry our wash. The warm, dry cellar is also an excellent storage space. We have only one bathroom in our home. This is sufficient for Lea, Terry, and me. The other three kids are used to their other home, which has four bathrooms. We could use an extra bathroom when they come for the weekend. Soon we will build a basic bathroom in the heated basement to be used while we are a family of six.

The best part of the change in our system for heating the house is that it did not become a compromise. We did not give up advantages that we already had. The change resulted in improvements to our home. We did not impact our electrical loads at all. The heating from the basement actually improved our balance of passive solar heat and wood heat. In the past, we ran the wood stove when the sun was not shining. We were sometimes fooled. Some days, after a cloudy start, the sun came out unexpectedly. Since we had already started the wood stove, we overheated the house. In the spring and the fall, it was difficult to warm the house and not overheat it. The stove installation in the cellar is a more forgiving match to our New England weather and solar heat. It is hard to quickly overheat the house from the cellar. The walls of the cellar are insulated on the outside. The suspended slab floors and the cellar walls create a giant heat sink. If the cellar gets too hot, it stores extra heat in the concrete masses and later gives this heat off slowly. We have maintained a balance in the interrelationship of our passive solar heat, our wood heat, and our solar electricity.

Our new cellar wood stove heats our house from the basement.

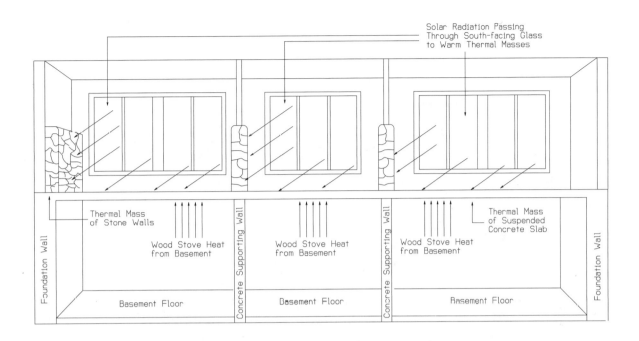

Solar Radiation Passing
Through South-facing Glass
to Warm Thermal Masses

Thermal Mass
of Stone Walls

Wood Stove Heat
from Basement

Thermal Mass
of Suspended
Concrete Slab

Wood Stove Heat
from Basement

Wood Stove Heat
from Basement

Foundation Wall

Concrete Supporting Wall

Concrete Supporting Wall

Foundation Wall

Basement Floor

Basement Floor

Basement Floor

**Our cellar wood stove
heats our suspended slab
from below at night and
on cloudy days while the
sun heats it from above
on suuny days.**

CHAPTER 26

What Is Next

IN 1989, we sold the 24 used batteries that composed our battery bank and replaced and expanded our battery bank with 32 new batteries. Our business was promoting a new brand of batteries, which we felt we should be displaying in our home system. The last year that Fowler Solar Electric Inc. was in our garage, we added eight 48 watt Hoxan modules to our solar electric system. The electrical energy demands of the business had increased. The year before, Lea and I had found ourselves conserving too much in the house to compensate for the increased use of the business. We installed the modules on the south side of the garage. The solar exposure at this location was diminished at certain times of the year by a young maple tree. Next summer we plan to build our porch off of the west end of our house. These modules will move to the south facing roof of the new porch, and the maple tree will continue to grow.

Our current solar electric system consists of a 32-battery, battery bank; 24 33-watt modules; and eight 48-watt modules. The battery bank is rated to store 42,000 watt-hours of electrical energy. The sum of all the modules is rated to produce 1,200 peak watts. This output is roughly equivalent to the output of 24 50-peak-watt modules that are sold today. Our 1,200-peak-watt arrays produce an average of 2,400 watt-hours per day of usable electrical energy in the winter months, and an average of 4,300 watt-hours per day of usable electrical energy in the summer months.

For most of the year we use our appliances, never worrying about the solar electric system. It fully charges the batteries on the next completely sunny day. We are careful with our electrical use during a span of four weeks between Thanksgiving and Christmas. Once every ten years we are confronted with a span of three weeks where we never have a sunny day. We still receive a small amount of output

The south side of our house and 24 solar electric modules.

from the modules in cloudy weather. We cut back our use as the length of a no-sun period increases. Our goal is to never discharge the battery bank below half full. Most years, we have a week of poor weather, then a sunny day that partially recharges the battery bank before the next cloudy week. We conserve only a small amount during these years.

I have lived with a solar electric system for twelve years. I ran a solar electric business for the majority of those years. For most of the years running the business, I spent most of each day designing solar electric systems and solving unique problems for solar-electric-system owners. Our system is ideal for our needs. Still, I confess that I have trouble leaving well enough alone. On any given day, I follow Lea around telling her my next daydream of improvements or future expansion. My rational side stops me from actually expanding our system. I do, however, have many ideas of what we will and will not do in the future.

I recently helped Bob and Karin Cook with the trial raising of a 64 foot wind tower, while Lea, Terry, and my stepson, Jarod, watched from a distance. In a few weeks the Cooks' 850 watt Bergey wind machine will arrive, and we will again raise the tower, this time with the new wind machine on top of it. Bob and Karin own a solar electric system that is about one third as large as ours. They have been constantly supplementing its smaller output with their large generator. Bob felt the output of the small solar electric system would balance well with the output of the wind machine.

I became instantly enthused about our neighbors' new wind machine. I borrowed Bob's information to study it. My first inclination was to install a wind machine this fall. Later, I decided to watch and wait while Bob and

We added eight, 48-watt modules on the garage which also charge our house battery bank. Two damaged and repaired modules that are mounted higher on the wall charge a 12V battery in the garage.

Karin live with their wind machine for the first year. I have many reservations about attaching a wind machine to our solar electric system. Now that I have had two months to daydream about a wind machine, I feel we will most likely never install one similar to the Cooks', even though I would love to watch it turn in the wind.

We do not need the output of an 850 watt wind machine. We could only utilize wind energy as a winter supplement to our large solar electric system. In theory, if we were to install a wind machine, we would only need a 250 watt wind machine. Whenever I add up the costs of even a small wind-machine installation, I quickly reach $1,500 to $2,000. I find that I can approximate the winter output of a wind machine investing the same amount of money on additional solar electric modules. I am confident that solar electric modules will last for over twenty years with absolutely no maintenance. I cannot imagine a rotating machine ever duplicating this kind of performance. I am even more certain that the wind machine will need maintenance when I begin to consider the dangers of lightning strikes or ice buildup on the propellers.

Lea and I became sure of our decision to not install a wind machine at the trial tower raising. Seeing the tower and guy wires in place provided us with a quick dose of a few realities. The first shock was the size of the area we would need to clear to install the guy wires. On paper I visualized guy wires disappearing into surrounding Christmas trees. After viewing the procedure of raising the tower, we were aware that we would really need to clear an area of more that 60 feet in all directions from the base of the tower. Our house is in a high lightning area. Lea is very worried about lightning strikes to the tower. We sometimes get the first lightning strike here with little warning. She worries for kids, running into the house under the tower and the guy wires, as a lightning bolt surges to ground. I have an additional personal worry. I am a tinkerer and worrier. I am capable of waking up in the night to go to the basement and check on our solar electric system. I like to double-check connections and test equipment for months after I install new components. Sometimes I get an idea later about a better way to install a part. These habits are incompatible with the installation of a wind machine on the top of a 60 foot tower. I am also aware that equipment fails at the worst times. For a New England wind machine, this would be winter. Bob's tower is designed to be lowered. However, I do not think I could lower a tower at my wind site if there were three feet of snow on the ground.

Karin jokingly advises me, "You'll probably need the extra electricity from a large wind machine when Terry insists on an electric guitar and amplifier." There is a lot of truth in the fact that things will change. My dream for a future wind machine is as follows: I want to someday build a 40 foot high observation tower to look out at the distant rolling hills. If wind machines become a popular addition to solar electric systems over the next few years, and smaller, high quality machines are available, I would like to install one on a 25 foot tower on top of my observation tower that would rise out of the 25 foot tall Christmas trees. I would install the machine after lightning season for the three winter months when we need supplemental power. The rest of the year it would sit safely in the garage.

I will probably never build my tower. There will probably always be more practical projects higher up on the list. Most likely, solar electric modules will gradually become cheaper, and I will buy more modules in lieu of a wind machine.

Our current 2,600 watt Trace inverter has a large internal battery charger that is meant to be powered by a 120VAC generator. If our need for power increases, the most logical addition to our alternative energy system would be

We expanded our bat-
tery bank in our house
basement from 24 batter-
ies to 32.

Bob and Karin Cook's wind machine installed on a 64 foot tower.

a portable generator to charge our battery bank during the worst winter months. We do not really want to invest in a generator, listen to its noise, or maintain another gasoline engine. However, this would be the next expansion that I would professionally recommend. If our loads increase, what we would really need would be the on-demand, winter supplement to our energy production. This could only be provided by a standby generator. Our choice for now would be to discharge our battery bank no lower than half full during a winter low-sun stretch, and to then rent a generator for a day to recharge our large battery bank to full.

We plan to replace our Trace 2600 watt inverter with the new Trace 4000 watt inverter in the next few months. The size of our current inverter is adequate. We will upgrade to the 4000 watt inverter for two reasons. This new inverter produces a true sine wave. Many appliances will run more efficiently. We will eliminate electrical interference noise in problem appliances and radio frequency interference on our AM radio reception and our telephones. The second reason we want the new Trace inverter is because this inverter has a much improved, and much larger, battery charger. If we ever need to rent a generator, we will have better and quicker capabilities for recharging the battery bank.

This summer as I considered a wind machine installation similar to the Cooks', I realize I have been dreaming of larger alternative energy systems that would power larger loads. On a small scale this is analogous to a utility company planning to build another power plant for customers whom they will have to convince to expand their use. If the power company would listen, I would advise them to urge their customers to conserve and avoid building that next power plant. On my small scale, I must also follow the same advice. In the future, I hope to harness my desire to design an expanded alternative energy system and direct it toward refining my conservation without changing my standard of living.

We currently use many less efficient light bulbs because they yield a more pleasing light, or because they fit neatly into specific fixtures. As our needs for more power increase, I hope to balance the increase by the substitution of compact fluorescent bulbs in certain lights especially for the one month of low energy production in the winter. We have other appliances, such as our stove vent fan, that run inefficiently on 120VAC, and I can retrofit them to run efficiently at 24VDC. Lea will soon be going to graduate school. I bought a used laptop for her last week at a tag sale. This computer uses 15 watts, while my desktop computer uses 75 watts. Perhaps I will replace my desktop with a notebook computer in the future. These few changes would reduce our daily electrical energy needs by the amount of electrical energy that five solar electric modules produce on an average winter day.

I always thought that I would build a batch solar hot-water heater when I sold Fowler Solar Electric Inc. I have not. Because we heat our water with an efficient tankless hot-water heater, it becomes less critical to invest the time and money to build and plumb a solar hot-water heater. Our tankless hot-water heater does have an optional regulator that will allow it to operate with preheated water input from another heating source. This winter, I will build a very simple system to preheat water from the heat of our wood stove.

Bob and Karin Cook are foremost our friends, and also our neighbors. Their homestead, energy systems, and way of life are more similar to ours than those of any other family in Worthington. One of us will try a new idea, or a new piece of equipment, and share the results with the other. Consequently, there are many similarities in our homesteads. There are also just as many differences created by sharing our individual successes and failures. Often a successful project for Bob and Karin will show Lea and me how well it works; at the same time, it shows us why we do not want it. An example is Karin's greenhouse. Lea and I watched each step of its construction and several years of use. The greenhouse they purchased and installed is well made. Watching the project convinced me that the foundation requirements, the labor, and the costs were more than I was willing to invest to have a greenhouse. Lea found that the maintenance and the difficulty of organically controlling infestations of aphids were not appealing to her. We chose not to have a greenhouse and to buy organic vegetables at our health-food-store supermarket during the non-gardening months.

In many ways, we use Bob and Karin's homestead as a means to develop our own ultimate direction. The part of the Cooks' homestead that we covet most is their large, beautiful swimming pond down in the woods. The pond sits in an open area that is beautifully maintained. We have only a small

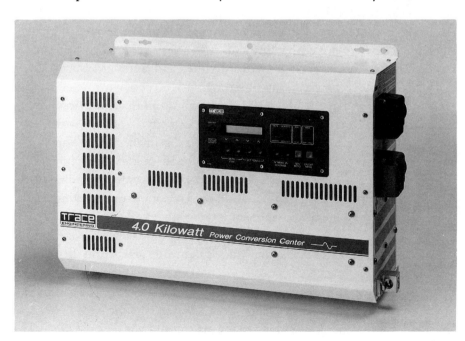

The new Trace 4,000 watt sine-wave inverter. Photo provided by Trace Engineering Company, Inc.

amount of land. Our original plot of land is nine acres. Across the road we own seven acres of wooded land. Our land has no suitable place to make a pond. Four years ago, when we sold the first half of our business to Steve, for the first time we had a chunk of money to put in our small retirement fund. One choice was to invest the money in a CD and read the monthly interest statements. We chose to invest the money in a seven-acre piece of land at a small private lake nearby. This gave us an investment that would grow and also one that would give us the pond we could not build.

I HATE TO MOW THE LAWN. There are always other more important and more peaceful chores to be done. Lea mows the lawn in our family. She grew up with neighbors in Montpelier, Vermont. I tease her that there is an inborn sense of suburban security related to the drone of a lawnmower. My environmentalist ways have always steered me to a philosophy that a large lawn rapes the countryside. I rebel against the attitude that we are man against nature, that the weeds and pests are always threatening, and that man must fight them off. Lea arrived and gradually expanded our lawn. Fortunately, the lawnmower is stored in the tractor bay behind the tractor. She can gain access to it only if I move the tractor for her, which sometimes requires a little bargaining. Her weedwhacker is electric. Its 100 foot cord limits Lea's range of destruction.

Our two-and-a-half-year-old son, Terry.

Though I joke about our different tendencies, Lea and I are in happy agreement about our grounds. We watch Bob and Karin spending long days maintaining the lawn, pond site, and fields of their larger homestead. We realize that if we expand our grounds and then maintain them, we will not have time to spend with the kids or to swim at the lake. As we age, we will be less likely to want to maintain extended grounds. We have decided to gradually replace much of our current grounds with areas of natural ground covers, flowers, and bushes such as myrtle, daylilies, and laurel. We want to live in a wild, low maintenance, beautiful setting.

Our life here on Bashan Hill is different from the normal American life. When the power is out in the town, we have electricity. When the winter is severe and cold, we are warm and our heating bill is unaffected. We are aware of the stresses and the inconsistencies of modern life. We try to make decisions that avoid some of these pitfalls and give us a more rewarding life. Buddhists profess that suffering is a result of expectations and wanting. Their solution for a life of contentment is the cessation of suffering. They believe we should see the world as a glass half full, not as a glass half empty. New Englanders remind us that the grass is not really greener on the other side of the fence.

It was very strange to return to live in Worthington in 1979. I was a local boy in some ways, and in many other ways, I was an outsider from Boston. Initially I was uncomfortable going to the general store or the post office. I was not sure how I could fit back into the community. Over a few years, Worthington again became my home and community. I have the best of both worlds. As a local boy, I am a member of that part of the community that you can only be born to. As an educated person from Boston, I am friends with the newcomers. I am in love with my town now that I have my two-and-a-half-year-old son. Terry goes everywhere with me. On a typical Saturday morning we take the trash and recyclables to the "disposal." My childhood friend, Linda Mason, who manages the "disposal," has a big greeting for Terry and me. We go to the library and Julie says, "Hi, Terry. How are you today? Would you like a graham cracker?" At the general store, Judy Fisk greets Terry and turns his nickel into a pretzel. Terry waves to everyone. He thinks the whole world knows him. In a few years Terry will go to the same grammar school that I attended. He may even get an old desk with my initials carved in it.

Lea and I, and Terry too, are content here on Bashan Hill. We enjoy making our own power and growing much of our own food. Our other children, Bethy, Kurt, and Jarod, who live primarily with their father in a large house, on an income many times greater than ours, sometimes ask in their adolescent confusion, "Are you and Mom poor?" I answer them, "We are rich. We have everything we could possibly want."

Lea, Terry, and I in our living room sunspace.

INDEX